Results and Problems in Cell Differentiation

52



*Series Editors*
Dietmar Richter, Henri Tiedge

Wolfgang Meyerhof • Ulrike Beisiegel •
Hans-Georg Joost
Editors

# Sensory and Metabolic Control of Energy Balance

Springer

*Editors*
Wolfgang Meyerhof
Department of Molecular Genetics
German Institute of Human Nutrition
Arthur-Scheunert-Allee 114-116
14558 Nuthetal
Germany
meyerhof@dife.de

Ulrike Beisiegel
Institute for Molecular Cell Biology
University Medical Center Hamburg-Eppendorf (UKE)
University of Hamburg
Martinistrasse 52
20246 Hamburg, Germany
beisiegel@uke.de

Hans-Georg Joost
Department of Pharmacology
German Institute of Human Nutrition
Arthur-Scheunert-Allee 114-116
14558 Nuthetal
Germany
joost@dife.de

*Series Editors*
Dietmar Richter
Center for Molecular Neurobiology
University Medical Center Hamburg-Eppendorf (UKE)
University of Hamburg
Martinistrasse 52
20246 Hamburg
Germany
richter@uke.uni-hamburg.de

Henri Tiedge
The Robert F. Furchgott Center for Neural and Behavioral Science
Department of Physiology and Pharmacology
Department of Neurology
SUNY Health Science Center at Brooklyn
Brooklyn, New York 11203
USA
htiedge@downstate.edu

ISBN 978-3-642-26518-1 ISBN 978-3-642-14426-4 (eBook)
DOI 10.1007/978-3-642-14426-4
Springer Dordrecht Heidelberg London New York



*Cover design:* WMXDesign GmbH, Heidelberg, Germany

Printed on acid-free paper

Springer is part of Springer Science+Business Media (www.springer.com)

# Preface

During the last two decades, the prevalence of obesity has dramatically increased in western and westernized societies. Its devastating health consequences include hypertension, cardiovascular diseases, or diabetes and make obesity the second leading cause of unnecessary deaths in the USA. As a consequence, obesity has a strong negative impact on the public health care systems. Recently emerging scientific insight has helped understanding obesity as a complex chronic disease with multiple causes. A multileveled gene–environment interaction appears to involve a substantial number of susceptibility genes, as well as associations with low physical activity levels and intake of high-calorie, low-cost, foods. Unfortunately, therapeutic options to prevent or cure this disease are extremely limited, posing an extraordinary challenge for today's biomedical research community.

Obesity results from imbalanced energy metabolism leading to lipid storage. Only detailed understanding of the multiple molecular underpinnings of energy metabolism can provide the basis for future therapeutic options. Numerous aspects of obesity are currently studied, including the essential role of neural and endocrine control circuits, adaptive responses of catabolic and anabolic pathways, metabolic fuel sensors, regulation of appetite and satiation, sensory information processing, transcriptional control of metabolic processes, and the endocrine role of adipose tissue. These studies are predominantly fuelled by basic research on mammalian models or clinical studies, but these findings were paralleled by important insights, which have emerged from studying invertebrate models.

The sense of taste is a vigilant gatekeeper of our body as it provides important information about substances in our diet, thereby influencing our decisions as to what to eat and drink. The gustatory system has been selected during evolution to detect nonvolatile nutritive and beneficial compounds, e.g., sweet and umami taste identifies the caloric content of food and the sensing of salt plays a crucial role in electrolyte homeostasis. In contrast, sour and bitter tastes have evolved as central warning signals against the ingestion of sour-tasting spoiled food or unripe fruits, as well as potentially toxic substances, including plant alkaloids and other environmental toxins. Consequently, taste perception is coupled to a certain hedonic tone, with sweet taste being generally perceived as most pleasant and bitter taste causing usually strong aversive reactions.

As important entities in metabolic regulation, taste receptors and their cells appear to be critical elements in efficient feedback control loops allowing dynamic regulations of the gustatory system in response to metabolic signals. For instance, in the mouse, sweet taste sensitivity (via the sweet taste receptor) is inversely correlated with circulating leptin levels, a major satiety factor mainly secreted by adipose tissue. In man, sweet sensitivity is coupled to diurnal variation in leptin levels, and leptin gene, as well as leptin receptor gene, polymorphisms modulate sweet taste sensitivity.

The role of taste receptor genes in gastrointestinal tissue as well as their dynamic regulation in gustatory and nongustatory tissues in response to metabolic cues establishes an entirely new and fast developing research field with impact on fuel sensing, metabolic control, and ingestive behavior. This volume reflects the recent scientific progress in the field of fuel sensing in the mouth, GI tract, and brain. It extends to the olfactory bulb as a metabolic sensor of endocannabinoids or brain insulin and glucose concentrations and the brain–gut endocrine axis. The unexpected finding of the association between autophagy and lipid metabolism in adipose tissue not only provides new insights in the regulation of these complex processes but also highlights possibly novel therapeutic approaches to tackle the metabolic syndrome. This volume also touches relevant novel molecular and cellular mechanisms regulating energy metabolism and causes and consequences of obesity, as well as the identification and functional characterization of obesity genes. Finally, the volume contains chapters illustrating the use of insect models to study relevant problems of energy homeostasis.

The idea for this book developed during the organization of the 30th Blankenese Conference, an international meeting that was sponsored by VolkswagenStiftung and held in May 2010 in Blankenese, a suburb of Hamburg, located in a picturesque setting above the river Elbe. This Conference acknowledged the aforementioned progress we have seen in the fields of sensory and metabolic regulation.

Postsdam, 2010
Hamburg, May 2010

Wolfgang Meyerhof, Postsdam
Ulrike Beisiegel, Hamburg
Hans-Georg Joost, Potsdam

# Contents

# The Genetic Basis of Obesity and Type 2 Diabetes: Lessons from the New Zealand Obese Mouse, a Polygenic Model of the Metabolic Syndrome

**Hans-Georg Joost**

**Abstract** The New Zealand obese (NZO) mouse is a polygenic model of severe obesity and type 2 diabetes-like hyperglycaemia. Outcross experiments with lean strains have led to the identification of numerous susceptibility loci (quantitative trait loci (QTL)) for adiposity and/or hyperglycaemia. Several major QTL were successfully introgressed into lean strains, and two responsible genes, the Rab-GAP *Tbc1d1* and the transcription factor *Zfp69*, were so far identified by a conventional strategy of positional cloning. *Tbc1d1* controls substrate utilization in muscle; SJL mice carry a loss-of-function variant that shifts substrate oxidation from glucose to fat and suppresses adiposity as well as development of diabetes. The zinc finger domain transcription factor *Zfp69* appears to regulate triglyceride storage in adipose tissue. Its normal allele *Zfp69* causes a redistribution of triglycerides from gonadal stores to liver, and consequently enhances diabetes when introgressed from SJL into NZO, whereas the loss-of-function variant present in NZO and C57BL/6J reduces the prevalence of diabetes. Data from human patients suggest that the orthologs of both genes may play a role in the pathogenesis of the human metabolic syndrome. In addition to *Tbc1d1* and *Zfp69*, variants of *Lepr*, *Pctp*, *Abcg1*, and *Nmur2* located in other QTL were identified as potential candidates by sequencing and functional studies. These results indicate that dissection of the genetic basis of obesity and diabetes in mouse models can identify novel regulatory mechanisms that are relevant for the human disease.

H.-G. Joost
German Institute of Human Nutrition Potsdam-Rehbrücke, Arthur-Scheunert-Allee 114-116, 14558 Nuthetal, Germany
e-mail: joost@dife.de

W. Meyerhof et al. (eds.), *Sensory and Metabolic Control of Energy Balance*, Results and Problems in Cell Differentiation 52,
DOI 10.1007/978-3-642-14426-4_1, 

# 1 Introduction

In previous years, mouse models have been shown to be of particular value for the elucidation of the pathogenesis of metabolic diseases such as obesity and diabetes. In 1994, the gene responsible for extreme obesity in the *ob*/*ob* mouse was cloned and shown to encode leptin, an anorexigenic peptide secreted from the adipocyte (Zhang et al. 1994). This landmark discovery was soon followed by identification of other obesity gene in monogenic mutant strains, namely the db (leptin receptor), agouti yellow, tubby, fat, and mahogany mutations (Friedman 1997; Leibel et al. 1997). These findings led to the elucidation of the neuro-endocrine regulation of hunger and satiety controlled by peptides such as MSH and NPY (Woods and D′Alessio 2008). Subsequently, it was shown that mutations of human orthologs (*LEP*, *LEPR*) or functionally associated genes (*POMC*, *MCR4*) produced phenotypes comparable with that of the respective mouse models (Clement et al. 1998; Santini et al. 2009).

## *1.1 Monogenic vs. Polygenic Traits*

Monogenic obesity is a very rare abnormality in humans as well as in mice, and it became soon apparent that genes causing monogenic obesity were not involved in the development of the common polygenic obesity. Furthermore, identification of the genes responsible for polygenic obesity in mice proved to be difficult and time consuming. Nevertheless, several groups tried to identify these genes, assuming that they would be informative for the human disease. The conventional strategy for identification of mouse disease genes is the genome-wide linkage analysis of outcross populations. This approach led to numerous susceptibility loci (quantitative trait loci, QTL) of obesity and diabetes-related traits such as body weight, fat mass, plasma glucose or insulin levels, β-cell mass [for review and meta-analysis of these studies see Wuschke et al. (2007) and Schmidt et al. (2008)]. However, so far only few responsible gene variants have been identified in these QTL.

## *1.2 The New Zealand Obese Mouse Model*

New Zealand obese (NZO) mice present a syndrome of morbid obesity, insulin resistance, hypertension, and hypercholesterolemia, which resembles the human metabolic syndrome (Bielschowsky and Bielschowsky 1953; Herberg and Coleman 1977; Ortlepp et al. 2000). Obesity develops in males and females within 4–8 weeks after weaning and can reach a body fat content of 40–50%. The positive energy balance is due to a moderately increased food intake and reduced energy expenditure with reduced body temperature (Koza et al. 2004; Jürgens et al. 2006). As a consequence of the syndrome, male NZO mice develop type 2-like diabetes characterized by marked hyperglycaemia, low serum insulin levels, and β-cell destruction (Leiter et al. 1998). The syndrome has a polygenic basis, and outcross progeny of the strain has previously been used for identification of QTL associated with adiposity, hypercholesterolemia, and hyperglycaemia (Leiter et al. 1998; Reifsnyder et al. 2000; Plum et al. 2000, 2002; Kluge et al. 2000; Taylor et al. 2001; Giesen et al. 2003; Vogel et al. 2009). By combining QTL from NZO and NON mice in different congenic lines, Reifsnyder and Leiter (2002) dissected the genetic interactions of the QTL and partially dissociated obesity and diabetes.

### *1.3 Consequences of Obesity on Glucose Homeostasis*

Obesity exerts different effects on glucose homeostasis in mice. Most, if not all, obese mice are insulin resistant and therefore more or less glucose intolerant. In some of these models, insulin resistance is compensated by hyperinsulinemia, and plasma glucose is only moderately elevated. Only a few models such as *db/db* and NZO, however, present overt diabetes mellitus as defined by a threshold of 16.6 mM (300 mg/dl) plasma glucose (Leiter et al. 1998). Mice crossing this threshold usually exhibit progressive β cell failure and subsequent β cell loss. In the C57BLKS-*db/db* strain, diabetes begins as early as at weeks 10–12 with a prevalence of 100%, whereas in other models such as NZO diabetes starts later (week 16–22) and reaches a prevalence of 40–60%. In NZO, onset and prevalence can be increased by feeding a high caloric, high-fat diet (Plum et al. 2002). In contrast to the *db/db* mutation (background strain C57BLKS), *ob/ob* (background strain C57BL/6J) failed to produce diabetes, glucose levels were normal because of massive β-cell proliferation and high serum insulin levels (Coleman 1978). This discrepancy led to the discovery of the diabetes-sensitive and diabetes-resistant backgrounds: introgression of the ob mutation into the BLKS (Coleman 1978) or BTBR background (Stoehr et al. 2000) resulted in diabetic lines, whereas the db allele introgressed into C57BL/6J failed to cause diabetes. Recently, two genes were identified that are presumably responsible for diabetes sensitivity of background strains: *Sorcs1* in BTBR mice (Clee et al. 2006) and *Lisch-like* in BLKS (Dokmanovic-Chouinard et al. 2008).

## 2 Identification of Gene Variants Responsible for Obesity and Diabetes in the NZO Mouse

Figure 1 depicts all significant and suggestive QTL we identified in crosses of NZO with lean strains. In order to identify the genes that are responsible for effects of individual QTL, we have employed three different strategies. First, we sequenced all candidate genes that were known to modify energy balance and were located in a QTL. By this approach, we identified variants of the leptin receptor (Igel et al. 1997)

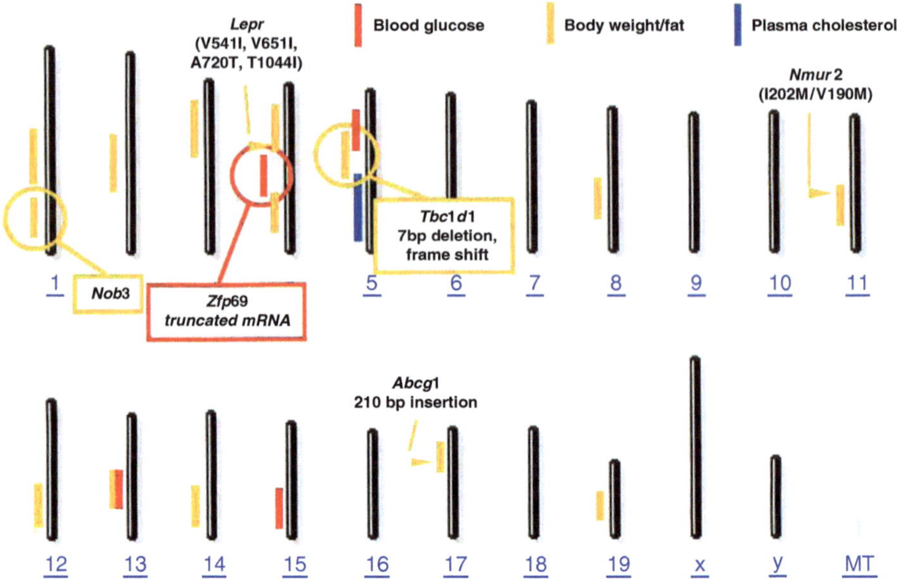


**Fig. 1** Chromosomal localization of QTL and candidate genes identified in genome-wide linkage analysis of outcross populations of NZO with SJL, NZB, and C57BL/6J. *Encircled loci* were successfully introgressed in a different strain

and the Nmu receptor 2 (Schmolz et al. 2007) that exhibited discrete functional alterations. Second, we took advantage of previous genome-wide mutagenesis (*Drosophila melanogaster*) and siRNA approaches (*Caenorhabditis elegans*). Mouse orthologs of genes that were found to be associated with adiposity in these studies, and that were located in one of the QTL, were sequenced and analyzed. By this approach, we identified a variant of the cholesterol transporter Abcg1 from NZO which increased adiposity (Buchmann et al. 2007). Third, we employed the conventional strategy of positional cloning and successfully introgressed three of the QTL (*Nob1*, *Nidd/SJL* and *Nob3*) into a different background (Chadt et al. 2008; Scherneck et al. 2009; Vogel et al. 2009). This strategy led to the identification of two candidate genes (*Tbc1d1* from *Nob1* and *Zfp69* from *Nidd/SJL*). Attempts to isolate a fourth QTL from chromosome 13 on the C57BL/6J background were not successful, because the resulting recombinant line showed no difference in adiposity as compared with controls (Vogel et al. unpublished).

## 2.1 *The Obesity Gene Tbc1d1 and Its Function as a Regulator of Substrate Oxidation in Skeletal Muscle*

By genome-wide linkage analysis of an outcross population of NZO with lean SJL mice, we identified a major QTL for body weight on chromosome 5 (Kluge et al. 2000).

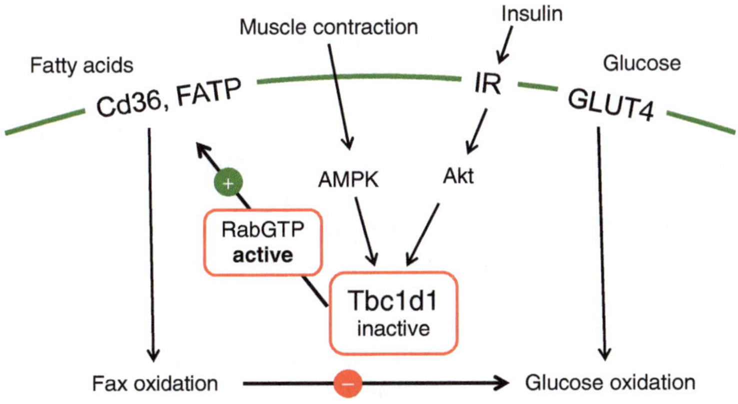


**Fig. 2** Presumed function of Tbc1d1 in the regulation of substrate utilization in muscle. When Tbc1d1 is inactive, RabGTP is not hydrolyzed, causing enhanced transport and oxidation of fatty acids which in turn reduces glucose transport

Additional outcross experiments with other lean strains suggested that the allele responsible for the weight difference was unique for SJL, and thus represented an obesity suppressor. Consequently, recombinant congenic lines were generated by introgression of the SJL allele of the QTL into C57BL/6J mice, and these lines were subsequently crossed with NZO in order to produce obesity. Characterization of the lines led to the definition of a critical region of the QTL that was characterized by sequencing and gene expression profiling with a custom-made array. By this approach, it was shown that SJL mice carry a loss-of-function variant of the RabGAP *Tbc1d1* generated by a 7 bp in-frame deletion, which produces a truncated protein (Chadt et al. 2008). Introgression of the variant into NZO reduced body weight and suppressed the development of diabetes. In C57BL/6J mice, the variant enhanced in vivo as well as ex vivo fat oxidation and reduced glucose oxidation. Similarly, knockdown of endogenous *Tbc1d1* in C2C12 muscle cells increased palmitate uptake and oxidation and reduced glucose oxidation (Chadt et al. 2008). In conclusion, adiposity and diabetes in obese mice are modified by disruption of *Tbc1d1* through a metabolic shift from glucose to fat oxidation (Fig. 2). This mechanism may also explain the previously reported association of a R125W variant with human obesity in Utah (Stone et al. 2006) and French families (Meyre et al. 2008).

## 2.2 *Zfp69, a Transcription Factor Associated with Altered Triglyceride Distribution and Enhanced Diabetes Susceptibility*

The genome-wide linkage analysis of a backcross population of NZO with SJL had shown that SJL contributed a diabetogenic allele located on distal chromosome 4

(*Nidd/SJL*). This QTL was responsible for acceleration and aggravation of the diabetes in the backcross population (Plum et al. 2000). The diabetogenic effect of the QTL was markedly enhanced by NZO chromosome 4 (*Tbc1d1*) and a high-fat diet (Plum et al. 2002): 70% of the mice carried both diabetogenic alleles (*Tbc1d1* and *Nidd/SJL*) but only 10% of the controls developed diabetes by week 23.

A critical interval of distal chromosome 4 (2.1 Mbp) conferring the diabetic phenotype was identified by interval-specific congenic introgression of SJL into diabetes-resistant C57BL/6J and by subsequent reporter cross with NZO. Analysis of the ten genes in the critical interval by sequencing and qRT-PCR revealed a striking allelic variance of the zinc finger domain transcription factor 69 (*Zfp69*); in NZO and C57BL/6J, mRNA of *Zfp69* was nearly undetectable. Analysis of the cDNA by 5′ and 3′RACE-PCR indicated a premature polyadenylation of the mRNA that was due to the presence of a retroviral transposon (IAPLTR1a) in intron 3 of the gene in NZO and C57BL/6J. The transposon disrupted the gene by the formation of a truncated mRNA that lacked the coding sequence for the KRAB and Znf-C2H2 domains of *Zfp69*. In contrast, the diabetogenic alleles from SJL, NON, and NZB lacked the transposon and generated a normal mRNA. When combined with the B6.V-$Lep^{ob}$ background, the diabetogenic $Zfp69^{SJL}$ allele produced hyperglycaemia, reduced gonadal fat, and increased plasma and liver triglycerides. mRNA levels of the human ortholog of *Zfp69*, *ZNF642*, were significantly increased in adipose tissue from patients with type 2 diabetes.

We conclude that *Zfp69* is the most likely candidate for the diabetogenic effect of *Nidd/SJL*. Expression of the transcription factor in adipose tissue may play a role in the pathogenesis of type 2 diabetes. In addition, retrotransposon IAPLTR1a appears to contribute substantially to the genetic heterogeneity of mouse strains, since we found that it produced aberrant mRNA species in seven other genes (Scherneck et al. 2009).

## 2.3 *Identification of Other Candidates Potentially Contributing to Obesity and Diabetes in NZO*

### 2.3.1 Leptin Receptor (*Lepr*)

It has been suggested that leptin resistance is a primary cause of the obesity in NZO mice (Halaas et al. 1997; Igel et al. 1997). Consistent with this finding, sequencing revealed that NZO carries a leptin receptor variant with four amino acid exchanges including two nonconservative substitutions (A720T and T1044I) (Igel et al. 1997). In order to assess the contribution of *Lepr* to the obesity syndrome of the NZO strain, female (SJL×NZO)NZO backcross mice were genotyped for the *Lepr* locus (Kluge et al. 2000). The variant allele appeared to enhance the effect of the QTL *Nob1* (wild-type *Tbc1d1*) on body weight. Females with homozygosity at both loci (*Lepr* N/N and *Tbc1d1* N/N) exhibited an average 22-week body weight of 60.6 g, whereas animals heterozygous at both loci (*Lepr* S/N and *Tbc1d1* S/N) showed an average 22-week body weight of only 51.9 g ($p = 0.0002$). Accordingly, homozygotes exhibited 2.2-fold higher serum insulin levels (35.7 ng/ml) than heterozygotes (16.0 ng/ml). Functional studies of the receptor variant expressed in COS-7 cell indicated only small reduction of its signaling potential (Kluge et al. 2000). Furthermore, the contribution of the receptor variant to the obesity syndrome of the NZO mouse appears to depend on other adipogenic alleles, since the variant is also present in the related New Zealand Black (NZB) strain, which shows neither obesity nor insulin resistance.

### 2.3.2 Phosphatidyl Choline Transfer Protein (*Pctp*)

It has been suggested that an abnormal phosphatidyl choline metabolism is involved in the development of diabetes in NZO mouse, since the hepatic activities of its two key enzymes were lower in NZO×NON mice than in the parental strains (Pan et al. 2005). Consequently, the genes involved in phosphatidyl choline metabolism and located in the QTL identified with this cross were further analyzed. Indeed, the phosphatidyl choline transfer protein (PC-TP) is located in the QTL *Nidd3*, which is responsible for hyperglycaemia and hypoinsulinemia. PC-TP is a specific phosphatidyl choline–binding protein and regulates hepatic lipid metabolism (Pan et al. 2006). Sequencing and functional studies indicated that NZO carries an inactive variant (R120H), which could be responsible for the diabetogenic effect of *Nidd3*.

### 2.3.3 ATP-binding Cassette Transporter G1 (*Abcg1*)

The ATP-binding cassette transporter G1 (Abcg1) catalyzes export of cellular cholesterol. Its association with obesity was discovered in a screening approach that compared data from *D. melanogaster* and mouse genomes (Buchmann et al. 2007): *Drosophila* genes involved in triglyceride storage were identified in a mutagenesis screen, and their mouse orthologs that are located in QTL were further analyzed. Overexpression of CG 17646, the *Drosophila* ortholog of *Abcg1*, generated lines of flies with increased triglyceride stores. *Abcg1* is located in a suggestive obesity QTL on proximal chromosome 17, and NZO mice carry a variant that was associated with higher expression of the gene in white adipose tissue, presumably because of an insertion containing multiple LXR-responsive elements. Targeted disruption of *Abcg1* in mice reduced adipose tissue depots, decreased the size of the adipocytes, and prevented high-fat diet-induced insulin resistance and fatty liver, corresponding with an increased expression of glucose transporter GLUT4, fatty acid transporter FATP1, and short chain 3-hydroxyacyl-*CoA*-dehydrogenase in adipose tissue. Furthermore, mRNA levels of the cholesterol-regulated transcription factor LXRα, and its downstream target *ABCA1*, were increased in adipose tissue of male $Abcg1^{-/-}$ mice. Thus, it has been suggested that ABCG1 regulates triglyceride storage by controlling intracellular cholesterol as a key regulator of gene expression in the adipocyte.

### 2.3.4 Neuromedin U Receptor 2 (*Nmur2*)

By sequencing of genes that are known to be involved in the regulation of energy balance, a variant of the neuromedin U receptor 2 ($Nmur2^{V190M/I202M}$) was identified (Schmolz et al. 2007). The *Nmur2* gene is located in a suggestive obesity QTL on chromosome 11, proximal to the diabetes QTL *Nidd3*. The receptor mediates a hypothalamic, anorexigenic effect of icv-administrated neuromedin on meal frequency. This effect was markedly reduced in NZO as compared with lean C57BL/6J mice (15% vs. 60% reduction in C57BL/6J). Transfection of HEK293 cells with wild-type *Nmur2* cDNA resulted in a dose-dependent calcium increase in response to a stimulation with NmU with an effective concentration ($EC_{50}$) of $3.0 \pm 1.3$ nM. In contrast, cells expressing the $NmuR2^{V190M/I202M}$ variant exhibited significantly ($p < 0.001$) higher $EC_{50}$ of $8.7 \pm 3.9$ nM. These data suggest that resistance to the anorexigenic effect of NmU contributes to the polygenic obesity of NZO mice and that it may reflect an impaired signal transduction of the $NmuR2^{V190M/I202M}$ variant (Schmolz et al. 2007).

## 3 Conclusions and Future Perspectives

The identification of the adipogenic/diabetogenic alleles of *Tbc1d1* and *Zfp69* supports the concept that fat oxidation and fat storage are crucial determinants of obesity and diabetes (Fig. 3). In addition, there is evidence that both genes can play

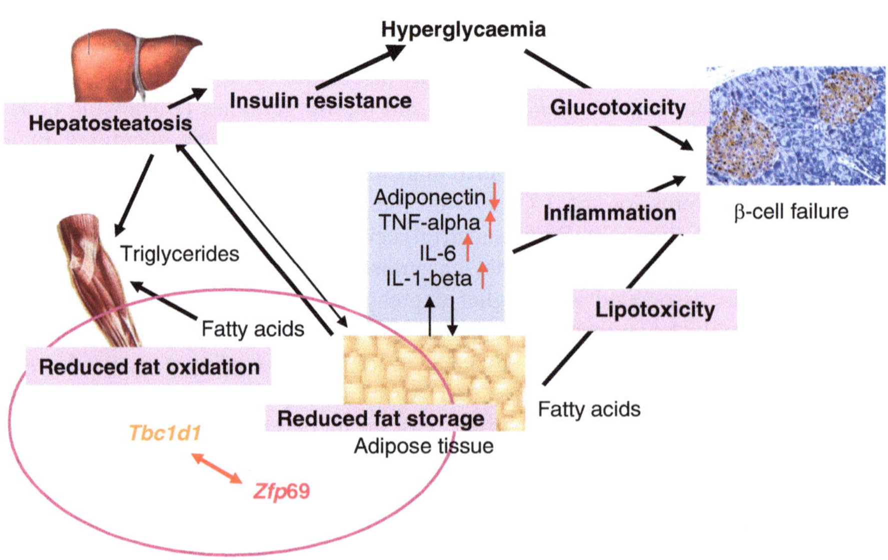


**Fig. 3** Pathogenesis of insulin resistance and β-cell failure. In this process, Tbc1d1 and Zfp69 play a major role by reducing fat oxidation and storage, respectively

a role in the human disease. Thus, the identification of mouse obesity and diabetes genes is a reasonable strategy to study the pathogenesis of both mouse and human disease.

In addition, the data have confirmed that conventional positional cloning is the preferable strategy to identify mouse genes that are responsible for a quantitative trait: By this strategy, small critical regions that harbor the gene in question can be identified with certainty. This strategy may not be applicable for minor QTL whose effects require the presence of other variants (epistasis) that are therefore lost upon introgression into another background. For identification of these genes, strategies for the screening of larger chromosomal intervals have to be developed.

The data also indicate that variants can be contributed by a presumed "healthy" strain; the obesity-suppressing, loss-of-function mutation of *Tbc1d1* was contributed by the lean SJL strain. Furthermore, the nondiabetic SJL strain contributed a diabetogenic allele (wild-type *Zfp69*). Loss-of-function mutations may reduce disease susceptibility rather than increase it, as was the case with $Tbc1d1^{SJL}$ and $Zfp69^{B6/NZO}$, where the "normal" alleles are adipogenic or diabetogenic, respectively.

**Acknowledgments** The work summarized in this paper was supported by grants from Deutsche Forschungsgemeinschaft, the European Union (Network of Excellence EUGENE2), and the German Ministry of Science and Technology (National Network of Genome Research, NGFN+, and German Centre of Diabetes Research, DZD).

## References

Bielschowsky M, Bielschowsky F (1953) A new strain of mice with hereditary obesity. Proc Univ Otago Med Sch 31:29–31

Buchmann J, Meyer C, Neschen S, Augustin R, Schmolz K, Kluge R, Al-Hasani H, Jürgens H, Eulenberg K, Wehr R, Dohrmann C, Joost HG, Schürmann A (2007) Ablation of the cholesterol transporter adenosine triphosphate-binding cassette transporter G1 reduces adipose cell size and protects against diet-induced obesity. Endocrinology 148:1561–73

Chadt A, Leicht K, Deshmukh A, Jiang LQ, Scherneck S, Bernhardt U, Dreja T, Vogel H, Schmolz K, Kluge R, Zierath JR, Hultschig C, Hoeben RC, Schürmann A, Joost HG, Al-Hasani H (2008) Tbc1d1 mutation in lean mouse strain confers leanness and protects from diet-induced obesity. Nat Genet 40:1354–1359

Clee SM, Yandell BS, Schueler KM, Rabaglia ME, Richards OC, Raines SM, Kabara EA, Klass DM, Mui ET, Stapleton DS, Gray-Keller MP, Young MB, Stoehr JP, Lan H, Boronenkov I, Raess PW, Flowers MT, Attie AD (2006) Positional cloning of Sorcs1, a type 2 diabetes quantitative trait locus. Nat Genet 38:688–693

Clement K, Vaisse C, Lahlou N, Cabrol S, Pelloux V, Cassuto D, Gourmelen M, Dina C, Chambaz J, Lacorte JM, Basdevant A, Bougneres P, Lebouc Y, Froguel P, Guy-Grand B (1998) A mutation in the human leptin receptor gene causes obesity and pituitary dysfunction. Nature 392:398–401

Coleman DL (1978) Obese and diabetes: two mutant genes causing diabetes-obesity syndromes in mice. Diabetologia 14:141–148

Crofford OB, Davis CK (1965) Growth characteristics, glucose tolerance and insulin sensitivity of New Zealand Obese mice. Metabolism 14:271–280

Dokmanovic-Chouinard M, Chung WK, Chevre JC, Watson E, Yonan J, Wiegand B, Bromberg Y, Wakae N, Wright CV, Overton J, Ghosh S, Sathe GM, Ammala CE, Brown KK, Ito R, LeDuc C, Solomon K, Fischer SG, Leibel RL (2008) Positional cloning of "Lisch-Like", a candidate modifier of susceptibility to type 2 diabetes in mice. PLoS Genet 4:e1000137

Friedman JM (1997) The alphabet of weight control. Nature 385:119–120

Giesen K, Plum L, Kluge R, Ortlepp J, Joost HG (2003) Diet-dependent obesity and hypercholesterolemia in the New Zealand obese mouse: identification of a quantitative trait locus for elevated serum cholesterol on the distal mouse chromosome 5. Biochem Biophys Res Commun 304:812–817

Halaas JL, Boozer C, Blair-West J, Fidahusein N, Denton DA, Friedman JM (1997) Physiological response to long-term peripheral and central leptin infusion in lean and obese mice. Proc Natl Acad Sci U S A 94:8878–8883

Herberg L, Coleman DL (1977) Laboratory animals exhibiting obesity and diabetes syndromes. Metabolism 26:59–98

Igel M, Becker W, Herberg L, Joost H-G (1997) Hyperleptinemia, leptin resistance and polymorphic leptin receptor in the New Zealand Obese (NZO) mouse. Endocrinology 138:4234–4239

Jürgens HS, Schürmann A, Kluge R, Ortmann S, Klaus S, Joost HG, Tschöp MH (2006) Hyperphagia, lower body temperature, and reduced running wheel activity precede development of morbid obesity in New Zealand obese mice. Physiol Genomics 25:234–241

Kluge R, Giesen K, Bahrenberg G, Plum L, Ortlepp JR, Joost HG (2000) Two quantitative trait loci for obesity and insulin resistance (*Nob1*, *Nob2*) and their interaction with the leptin receptor locus ($Lepr^{A720T/T1044I}$) in New Zealand obese (NZO) mice. Diabetologia 43:1565–1573

Koza RA, Flurkey K, Graunke DM, Braun C, Pan HJ, Reifsnyder PC, Kozak LP, Leiter EH (2004) Contributions of dysregulated energy metabolism to type 2 diabetes development in NZO/H1Lt mice with polygenic obesity. Metabolism 53:799–808

Leibel RL, Chung WK, Chua SC Jr (1997) The molecular genetics of rodent single gene obesities. J Biol Chem 272:1937–1940

Leiter EH, Reifsnyder PC, Flurkey K, Partke H-J, Junger E, Herberg L (1998) NIDDM genes in mice. Deleterious synergism by both parental genomes contributes to diabetic thresholds. Diabetes 47:1287–1295

Meyre D, Farge M, Lecoeur C, Proenca C, Durand E, Allegaert F, Tichet J, Marre M, Balkau B, Weill J, Delplanque J, Froguel P (2008) R125W coding variant in TBC1D1 confers risk for familial obesity and contributes to linkage on chromosome 4p14 in the French population. Hum Mol Genet 17:1798–1802

Ortlepp JR, Kluge R, Giesen K, Plum L, Radke P, Hanrath P, Joost HG (2000) A metabolic syndrome of hypertension, hyperinsulinemia, and hypercholesterolemia in the New Zealand Obese (NZO) mouse. Eur J Clin Invest 30:195–202

Pan HJ, Agate DS, King BL, Wu MK, Roderick SL, Leiter EH, Cohen DE (2006) A polymorphism in New Zealand inbred mouse strains that inactivates phosphatidylcholine transfer protein. FEBS Lett 580:5953–5958

Pan HJ, Reifsnyder P, Vance DE, Xiao Q, Leiter EH (2005) Pharmacogenetic analysis of rosiglitazone-induced hepatosteatosis in new mouse models of type 2 diabetes. Diabetes 54:1854–1862

Plum L, Kluge R, Giesen K, Altmüller J, Ortlepp JR, Joost H-G (2000) Type-2-diabetes-like hyperglycemia in a backcross model of New Zealand obese (NZO) and SJL mice: characterization of a susceptibility locus on chromosome 4 and its relation with obesity. Diabetes 49:1590–1596

Plum L, Giesen K, Kluge R, Junger E, Linnartz K, Schürmann A, Becker W, Joost H-G (2002) Characterization of the diabetes susceptibility locus *Nidd/SJL* in the New Zealand obese (NZO) mouse: islet cell destruction, interaction with the obesity QTL *Nob1*, and effect of dietary fat. Diabetologia 45:823–830

Reifsnyder PC, Churchill G, Leiter EH (2000) Maternal environment and genotype interact to establish diabesity in mice. Genome Res 10:1568–1578

Reifsnyder PC, Leiter EH (2002) Deconstructing and reconstructing obesity-induced diabetes (diabesity) in mice. Diabetes 51:825–832

Santini F, Maffei M, Pelosini C, Salvetti G, Scartabelli G, Pinchera A (2009) Melanocortin-4 receptor mutations in obesity. Adv Clin Chem 48:95–109

Scherneck S, Nestler M, Vogel H, Blüher M, Block MD, Berriel Diaz M, Herzig S, Schulz N, Teichert M, Tischer S, Al-Hasani H, Kluge R, Schürmann A, Joost HG (2009) Positional cloning of zinc finger domain transcription factor Zfp69, a candidate gene for obesity-associated diabetes contributed by mouse locus Nidd/SJL. PloS Genet 5:e1000541

Schmidt C, Gonzaludo NP, Strunk S, Dahm S, Schuchhard J, Kleinjung F, Wuschke S, Joost HG, Al-Hasani H (2008) A metaanalysis of QTL for diabetes related traits in rodents. Physiol Genomics 34:42–53

Schmolz K, Pyrski M, Bufe B, Vogel H, Nogueras R, Jürgens H, Nestler M, Zahn C, Tschöp M, Meyerhof W, Joost HG, Schürmann A (2007) Regulation of feeding behavior in normal and obese mice by neuromedin-U: a variant of the neuromedin-U receptor 2 contributes to hyperphagia in the New-Zealand obese mouse. Obe Metab 3:28–37

Stoehr JP, Nadler ST, Schueler KL, Rabaglia ME, Yandell BS, Metz SA, Attie AD (2000) Genetic obesity unmasks nonlinear interactions between murine type 2 diabetes susceptibility loci. Diabetes 49:1946–1954

Stone S, Abkevich V, Russell DL, Riley R, Timms K, Tran T, Trem D, Frank D, Jammulapati S, Neff CD, Iliev D, Gress R, He G, Frech GC, Adams TD, Skolnick MH, Lanchbury JS, Gutin A, Hunt SC, Shattuck D (2006) TBC1D1 is a candidate for a severe obesity gene and evidence for a gene/gene interaction in obesity predisposition. Hum Mol Genet 15:2709–2720

Taylor BA, Wnek C, Schroeder D, Phillips SJ (2001) Multiple obesity QTLs identified in an intercross between the NZO (New Zealand obese) and the SM (small) mouse strains. Mamm Genome 12:95–103

Vogel H, Nestler M, Rüschendorf F, Block MD, Tischer S, Kluge R, Schürmann A, Joost HG, Scherneck S (2009) Characterization of Nob3, a major quantitative trait locus for obesity and hyperglycemia on mouse chromosome 1. Physiol Genomics 38:226–232

Woods SC, D'Alessio DA (2008) Central control of body weight and appetite. J Clin Endocrinol Metab 93(Suppl 1):S37–S50

Wuschke S, Dahm S, Schmidt C, Joost HG, Al-Hasani H (2007) A metaanalysis of QTL associated with body weight and adiposity. Int J Obes 31:829–841

Zhang Y, Proenca R, Maffei M, Barone M, Leipols L, Friedman JM (1994) Positional cloning of the mouse obese gene and its human homologue. Nature 372:425–432

# Regulation of Nutrient Metabolism and Inflammation

**Sander Kersten**

**Abstract** Metabolic and immune-related pathways intersect at numerous levels. Their common regulation is effectuated by several hormonal signaling routes that involve specific nuclear hormone receptors and adipokines. Glucocorticoids and leptin are hormones that play a key role in coordinating energy metabolism and food-seeking behavior during energy deficiency as does the nuclear hormone receptor Peroxisome Proliferator Activated Receptor $\alpha$ (PPAR$\alpha$). Importantly, the glucocorticoid, leptin, and PPAR$\alpha$ signaling routes share a profound role in governing inflammation and other immune-related processes. Using specific examples, this chapter aims at illustrating the interplay between metabolism and immunity/inflammation by discussing common endocrine and transcriptional regulators of metabolism and inflammation and by highlighting the interaction between macrophages and metabolically active cells in liver and adipose tissue. Convergence of metabolic and immune signaling is likely at least partially driven by the evolutionary need during times of food insufficiency to minimize loss of energy to processes that are temporarily nonessential to the survival of the species.

## 1 General Introduction

The present-day abundance of highly palatable foods combined with the astounding number of people suffering from diseases of affluence makes it hard to imagine that mankind evolved under less favorable nutritional circumstances. Most of our ancestors had to cope with repeated and prolonged periods of caloric insufficiency or even starvation, especially during winter or when traveling long distances. In fact, even in western societies, general food security was only attained two to

S. Kersten
Nutrition, Metabolism and Genomics Group, Division of Human Nutrition, Wageningen University, Bomenweg 2, 6703 HD Wageningen, The Netherlands
e-mail: sander.kersten@wur.nl

W. Meyerhof et al. (eds.), *Sensory and Metabolic Control of Energy Balance*, Results and Problems in Cell Differentiation 52,
DOI 10.1007/978-3-642-14426-4_2, © Springer-Verlag Berlin Heidelberg 2010

three generations ago, and undernutrition is still common in many parts of the world. As a consequence, caloric insufficiency and starvation have served as key evolutionary pressures throughout history that have shaped energy metabolism in humans and determined our metabolic response to the current food excess (Prentice et al. 2008).

In addition to food insufficiency, our ancestors were subjected to numerous infectious challenges, ranging from pathogenic bacteria to viruses and parasites, which triggered the evolution of a highly complex immune defense system that is able to ward off all but the most formidable pathogens. But although an efficiently working immune system was critical for the long-term survival of the human species, it was likely of lesser importance during period of energy shortage, when energy conservation must prevail. In Drosophila, immunodeficient mutant flies that do not have to invest into immunological maintenance were found to have an extended lifespan under starvation conditions compared to wild-type flies, whereas the opposite occurred under fed conditions (Valtonen et al. 2010). The pressure to conserve energy may explain why starvation and undernutrition can wreak havoc on the immune system and give rise to general immunosuppression.

But the interaction between metabolism and immunity is not just a one-way street. Indeed, infections are well recognized to have a major impact on nutrient metabolism, increasing energy requirements and, depending on the severity of infection, cause significant protein catabolism and cachexia (Goldstein and Elwyn 1989). Given the important links between metabolism and immunity, it comes as no surprise that the regulatory mechanisms that govern metabolism and immunity intersect at numerous levels. In fact, many of the endocrine factors that are responsible for coordinating metabolic pathways also directly govern inflammatory routes. Conversely, a growing list of circulating inflammatory and immune modulators has been shown to impact metabolic pathways. This chapter is aimed at illustrating the interplay between metabolism and immunity/inflammation by discussing common endocrine and transcriptional regulators of metabolism and inflammation and by highlighting the interaction between macrophages and metabolically active cells in liver and adipose tissue. The chapter does not aim to be exhaustive but rather uses specific examples to emphasize the importance of cross-talk between metabolic and immune regulation in response to disturbances of homeostasis.

## 2 Common Endocrine and Transcriptional Regulators of Metabolism and Inflammation

### *2.1 Regulation of Metabolism and Immunity by Leptin*

Although energy conservation is of prime importance to any organism at risk of experiencing major food scarcity to assure the survival of the species, the need to save energy becomes especially important during actual episodes of caloric insufficiency.

Under those circumstances, limiting energy usage and shifting metabolism toward oxidation of stored nutrients take precedence, while nonessential processes such as the immune response become deprioritized. The parallel changes in energy metabolism and the immune system are effectuated by a number of hormonal systems, three of which are discussed in more detail below.

One major hormone that aptly embodies the convergence of metabolic and immunological regulation is leptin (Fig. 1). Leptin is a 16-kDa protein predominantly secreted by adipocytes (Zhang et al. 1994). Consequently, plasma leptin levels are positively correlated with body fat mass. Moreover, plasma leptin levels acutely decrease during fasting, which via the leptin receptor OB-Rb triggers neuronal circuits in the arcuate nucleus of the hypothalamus (Friedman and Halaas 1998). Specifically, leptin increases production of satiety-inducing signals α Melanocyte-Stimulating Hormone and Cocaine- and Amphetamine-Regulated Transcript and decreases production of hunger-provoking peptides Neuropeptide Y and Agouti-Related Protein. Absence of leptin causes severe hyperphagia in mouse and humans, which combined with reduced energy expenditure leads to morbid obesity in both species (Farooqi and O'Rahilly 2009). Leptin increases energy expenditure in part by stimulating uncoupling in white and brown adipose tissue (Commins et al. 1999), as well as via increased spontaneous activity (Pelleymounter et al. 1995). The metabolic effects of leptin appear to be mediated by Adenosine MonoPhosphate-dependent kinase and by the signal transduction cascades that are also involved in the immune-modulating actions of leptin (Minokoshi et al. 2002).

Illustrating the close ties with immune function, the receptor for leptin is a member of the class I cytokine receptor family and is homologous to gp-130, the signaling transducing subunit of IL-6 family cytokines (Tartaglia et al. 1995). In addition to the hypothalamus, the leptin receptor can be found on numerous types of immune cells including subpopulations of T cells, B cells, dendritic cells, monocytes, neutrophils, macrophages, and natural killer cells (Fig. 1) (Gainsford

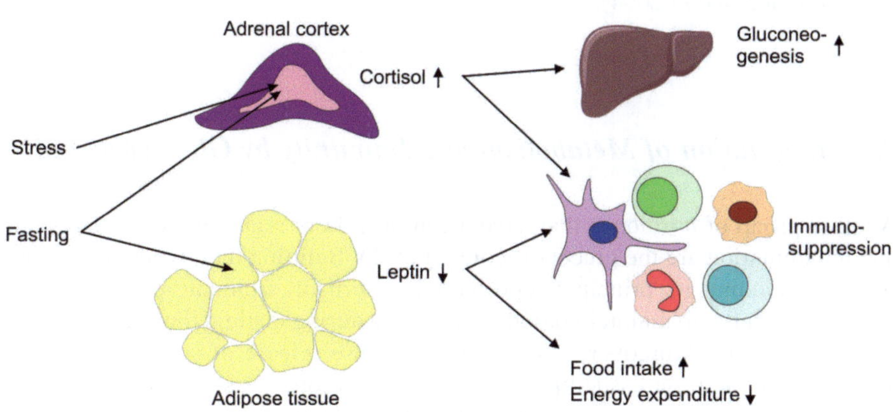


**Fig. 1** Importance of glucocorticoid and leptin signaling in coordinating metabolic and immune-related responses during stress and fasting

et al. 1996). Binding of leptin to the receptor and subsequent activation of downstream effector JAK2 causes phosphorylation and concurrent activation of STAT transcription factors, which are also targeted by other cytokines (Ghilardi et al. 1996). STAT proteins play a central role in the immune response by regulating the expression of numerous cytokines and other immunologically relevant genes. Phosphorylated JAK2 also activates other pathways including Mitogen-Activated Protein kinase/Extracellular signal-Regulated Kinase (ERK) and Phosphatidyl-Inositol 3-Kinase/Protein Kinase B. The latter pathway represents a key signaling cascade that mediates effects of various proinflammatory cytokines and bacterial and viral stimuli in a variety of immune cells. Through these mechanisms, leptin targets both innate and acquired immunity. With respect to innate immunity, leptin has been shown to enhance phagocytosis by monocyte/macrophages. Moreover, leptin stimulates secretion of proinflammatory mediators of the acute-phase response and the expression of adhesion molecules (La Cava and Matarese 2004). With respect to acquired immunity, a wealth of data is available showing that leptin impacts diverse aspects of T cell-mediated immunity, as first revealed by the higher number of circulating regulatory T cells in mice lacking leptin. Strikingly, the alterations in immune function in mice lacking leptin resemble the state of immunodeficiency observed with caloric insufficiency and starvation (Howard et al. 1999), suggesting that decreased leptin is a key mediator of the immune suppression of malnutrition (Lord et al. 1998).

Leptin can thus be regarded as a primordial hunger/satiety hormone that responds to acute and chronic energy shortage by promoting food-seeking behavior and energy conservation. It is likely that its immune-modulating function was acquired during the course of evolution as a mechanism to minimize energy losses via redundant processes. The alternative scenario in which leptin initially functioned as immune modulator that later adopted the ability to govern food intake seems much less plausible. Interestingly, similar convergence of metabolic and immune regulation is found for other adipokines such as adiponectin and resistin (Trayhurn and Wood 2004).

## 2.2 Regulation of Metabolism and Immunity by Glucocorticoids

Another group of hormones that have a central role in regulation of metabolism and inflammation are the glucocorticoids (Fig. 1). Befitting their name, glucocorticoids, including its principal representative cortisol, stimulate processes that collectively serve to maintain blood sugar levels during fasting, thus opposing the action of insulin. Moreover, glucocorticoids cause release of stored energy via stimulation of adipose tissue lipolysis and muscle proteolysis. As a consequence, pathologically increased circulating cortisol concentration 2 give rise to dysmetabolic features such central obesity, insulin resistance, and dyslipidemia (Mattsson and Olsson 2007). Apart from its role in nutrient metabolism, glucocorticoids have a major impact on inflammatory and immune-related processes (Cupps and Fauci 1982). Oral glucocorticoids represent the most widely used treatment for a great variety of acute and chronic inflammatory disorders, including chronic obstructive pulmonary diseases, inflammatory skin disorders, and rheumatoid arthritis. At the cellular level, glucocorticoids inhibit the access of leukocytes to inflammatory sites, interfere with the functions of leukocytes, endothelial cells, and fibroblasts, and suppress the production and the effects of humoral factors involved in the inflammatory and immune response (Boumpas et al. 1993). The pleiotropic effects of glucocorticoids, which extend far beyond inflammation and metabolism and also include effects on fetal development, brain function, and bone formation, are reflected in the diverse genes and processes controlled by the glucocorticoid receptor (GR), a member of nuclear receptors superfamily (Beck et al. 2009). While normally residing in the cytosol, GR translocates into the nucleus upon binding of glucocorticoids. There, GR serves as transcriptional activator of distinct glucocorticoid-responsive target genes via direct DNA binding to GR-responsive elements (GRE), in conformance with the general mechanism of gene regulation by nuclear receptors. Absence of GR in liver leads to fasting hypoglycemia, illustrating the importance of GR in maintaining blood glucose concentrations via stimulation of hepatic gluconeogenesis and its key targets phosphoenolpyruvate carboxykinase and glucose 6-phosphatase (Imai et al. 1990; Opherk et al. 2004). With respect to lipid metabolism, glucocorticoids have been shown to raise plasma-free fatty acid levels, reflecting increased whole body lipolysis, yet promote central fat deposition (Macfarlane et al. 2008).

Whereas glucocorticoids influence nutrient metabolism primarily by directly stimulating gene transcription, the anti-inflammatory and immune-suppressive effects of activated GR are mainly mediated by altering the activity of other DNA-bound transcription factors via a mechanism that does not require DNA binding of GR referred to as transrepression. Indeed, using mice with a mutation in the GR gene that prevents transcriptional activation of classical GRE-regulated genes, it was shown that repression of inflammatory responses by glucocorticoids is independent of GR binding to DNA (Reichardt et al. 2001). Many transcription factors are targeted for transrepression by GR, including NF-κB, AP-1, NF-AT, T-bet, GATA-3, and IRF3, thereby explaining the diverse effects of glucocorticoids on inflammation and immunity observed in a variety of different cell types. GR represses NF-κB transcriptional activity by physically interacting with the p65 subunit, preventing it from gaining access to DNA. Additionally, GR and NF-κB may compete for mutual cofactors required for transcriptional activation (Smoak and Cidlowski 2004).

Similar to the situation for leptin, the question can be raised why regulation of metabolism and inflammation is effectuated by a single hormone via a single receptor. Glucocorticoids allow the body to respond adequately to physical or emotional stress via adjustments in the metabolic flux to support those processes that are essential for the immediate survival of the organism. Although this may be the evolutionary basis for the comprehensive properties of GR, it seems plausible that additional immune-modulating functions of GR have evolved that are not strictly connected to the starvation response.

## 2.3 Regulation of Metabolism and Immunity by PPARα

Analogous to GR, another nuclear receptor governing both metabolic and immune-related processes is PPARα, which serves as the molecular target for hypolipidemic fibrate drugs. PPARα can be considered the master regulator of fatty acid metabolism in liver. It is especially active in the fasted state, giving rise to induction of numerous genes involved in fatty acid uptake, fatty acid oxidation, triglyceride synthesis and hydrolysis, and other fatty acid metabolic pathways (Kersten et al. 2000). Additionally, activation of PPARα has a major impact on inflammatory processes in liver, which is mainly accomplished via downregulation of gene expression by interfering with specific inflammatory signaling pathways, leading to attenuation of cytokine-induced production of acute-phase proteins such as fibrinogen, haptoglobin, and serum amyloid (Gervois et al. 2004; Mansouri et al. 2008). Similarly, the presence of PPARα protects against obesity-induced hepatic inflammation (Stienstra et al. 2007). With respect to the mechanisms involved, it has been shown that activated PPARα binds to c-Jun and the p65 subunit of NF-κB, thereby inhibiting AP-1- and NF-κB-mediated signaling (Delerive et al. 2001). Recent high throughput DNA binding studies of PPARα via ChIP on CHIP revealed that PPARα activation in hepatocytes causes the dissociation of STAT transcription factors from the DNA, resulting in downregulation of gene expression (van der Meer et al. 2010). As mentioned above, STATs are also targeted by leptin, suggesting potential synergy between the PPARα and leptin pathways to suppress STAT-dependent inflammatory signaling during fasting. In general, the mechanisms underlying the anti-inflammatory effects of PPARα are reminiscent to those of GR, and indeed cooperation between anti-inflammatory activities of the two receptors has been demonstrated (Bougarne et al. 2009).

Although synthetic fibrates potently activate PPARα to cause suppression of inflammation in liver and the vascular wall, it should be stressed that the receptor did not evolve as a drug target but rather as a lipid target with a preference for long chain unsaturated fatty acids and its derivatives. Numerous studies using a variety of biochemical techniques have firmly corroborated direct physical association between fatty acids and PPARs and have established fatty acids as bona fide PPAR ligands (Jump et al. 2005). Recently, it was shown that the effects of dietary unsaturated fatty acids on hepatic gene expression are almost entirely mediated by PPARα (Sanderson et al. 2008). PPARα may thus serve as mediator of the inflammatory activity of dietary fatty acids and thereby link diet composition to regulation of the inflammatory response in liver. A similar scenario can be envisioned for PPARβ/δ and PPARγ, which also function as lipid sensors and have major anti-inflammatory properties in several cells and tissues (Bishop-Bailey and Bystrom 2009; Straus and Glass 2007).

# 3 Interaction Between Macrophages and Metabolically Active Cells in Liver and Adipose Tissue

While there are thus numerous examples for coordinated regulation of metabolism and inflammation/immunity as part of normal homeostasis, cross-talk between metabolism and inflammation can also be triggered by specific disease conditions representing perturbations of homeostasis. Indeed, deviations of metabolic homeostasis as in obesity elicits marked changes in inflammatory pathways, and vice versa a strong inflammatory response can have a major impact on numerous metabolic processes, including lipoprotein metabolism, insulin signaling, and energy expenditure. Below, it is discussed how in the state of obesity interactions exist in liver and adipose tissue between macrophages and the metabolically active main cell types, which importantly contributes to some of the complications of obesity.

## *3.1 Role of Adipose Tissue Macrophages in Obesity*

Contradicting the widespread connotation of body fat with inactivity, over the last two decades, it has become evident that the adipose tissue is anything but inactive and represents a highly dynamic organ that rapidly adapts to changes in nutritional status via alterations in secretory profile, blood flow, and cell composition. Traditionally viewed as an organ composed almost entirely of adipocytes complemented with some vascular cells and fibroblasts, it is now commonly accepted that numerous types of leukocytes are present in between the adipocytes (Anderson et al. 2010). This pool of white blood cells expands dramatically upon obesity, creating an optimal forum for exchange of signals between immune cells and metabolically active cells. Most of the attention has been focused on macrophages, which were shown to infiltrate the adipose tissue and account for production of specific cytokines (Weisberg et al. 2003; Xu et al. 2003). In addition, adipocytes themselves are capable of secreting a variety of proteins, including TNF$\alpha$ and leptin (Hotamisligil et al. 1993; Zhang et al. 1994). During adipose tissue expansion, changes in the secretory profile of adipocytes and incoming and resident leukocytes lead to altered release of numerous (adipo)cytokine. Accordingly, obesity is nowadays considered a state of chronic low-grade inflammation, illustrating the intimate relationship between metabolism and inflammation.

How obesity leads to enhanced macrophage abundance is not clear but a role for adipocyte cell death, hypoxia, and local insulin resistance has been surmised, the latter being possibly related to ER stress, ROS formation, and mitochondrial dysfunction (Lee et al. 2009). A number of chemotactic signals including Monocyte Chemoattractant Protein 1 likely contribute to macrophage recruitment in adipose tissue (Kamei et al. 2006; Kanda et al. 2006; Yu et al. 2006). Recently, it was shown that T cells are also actively regulated in adipose tissue and contribute to obesity-induced inflammation (Feuerer et al. 2009; Nishimura et al. 2009; Winer et al. 2009).

Studies have also indicated that macrophages present within adipose tissue display pronounced heterogeneity and can roughly be separated into classically activated proinflammatory M1 macrophages and alternatively activated anti-inflammatory M2 macrophages. While adipose tissue of lean mice mainly contains M2 macrophages, diet-induced obesity was shown to specifically increase adipose abundance of M1 macrophages via recruitment from the circulation (Lumeng et al. 2008). The activation state of adipose tissue macrophages may in part be determined by T cells via specific T helper type 1 or T helper type 2 cytokines (Strissel et al. 2010). M1 macrophages are thought to contribute to development of insulin resistance in obese animals via production of proinflammatory cytokines, which negatively impact peripheral and hepatic insulin signaling. These effects are mediated by specific kinases such as Inhibitor of NF-κB Kinase β (IKKβ), c-Jun *N*-terminal Kinase (cJNK), ERK, mammalian Target Of Rapamycin (mTOR), and p70 ribosomal S6 protein Kinase (S6K) that cause phosphorylation of insulin receptor substrate 1 and 2 at specific serine residues, leading to their inactivation (Shoelson et al. 2007). Normally, binding of insulin to its receptor triggers phosphorylation of insulin receptor substrates at specific tyrosine residues, leading to their activation. In recent years, this so-called inflammatory hypothesis has gained acceptance as an at least partial explanation for the link between obesity and local and possibly systemic insulin resistance (Lee et al. 2009).

## *3.2 Cross-talk Between Hepatocytes and Kupffer Cells*

The close proximity and the chemical interaction between leukocytes and metabolically active cells is not an exclusive phenomenon of the obese adipose tissue. The liver represents a unique system in which the metabolically relevant hepatocytes coexist with resident macrophages referred to as Kupffer cells (Fig. 2). Kupffer cells are derived from circulating monocytes that arise from bone marrow progenitors. Once in the liver, the cells differentiate to Kupffer cells and acquire the ability to perform various functions, including phagocytosis, antigen processing, and antigen presentation. Via the generation of different products, including cytokines, prostanoids, nitric oxide, and reactive oxygen intermediates, Kupffer cells influence the phenotypes of other immune cells and neighboring hepatocytes. Recent studies point to important cross-talk between Kupffer cells and hepatocytes in the regulation of lipid metabolism in the context of nonalcohol fatty liver disease (NAFLD). NAFLD represents the most common liver disorder in industrialized countries and is highly correlated with obesity. It is characterized by accumulation of fat in hepatocytes and can vary from the relatively benign hepatic steatosis to severe steatohepatitis, which might progress to end-stage liver diseases such as cirrhosis and liver cancer. Several studies (Huang et al. 2010; Neyrinck et al. 2009; Rivera et al. 2007; Stienstra et al. 2010), though not all (Clementi et al. 2009; Lanthier et al. 2009), found that when liver was depleted of Kupffer cells using clodronate liposomes or gadolinium chloride, hepatic lipid accumulation in mouse models of

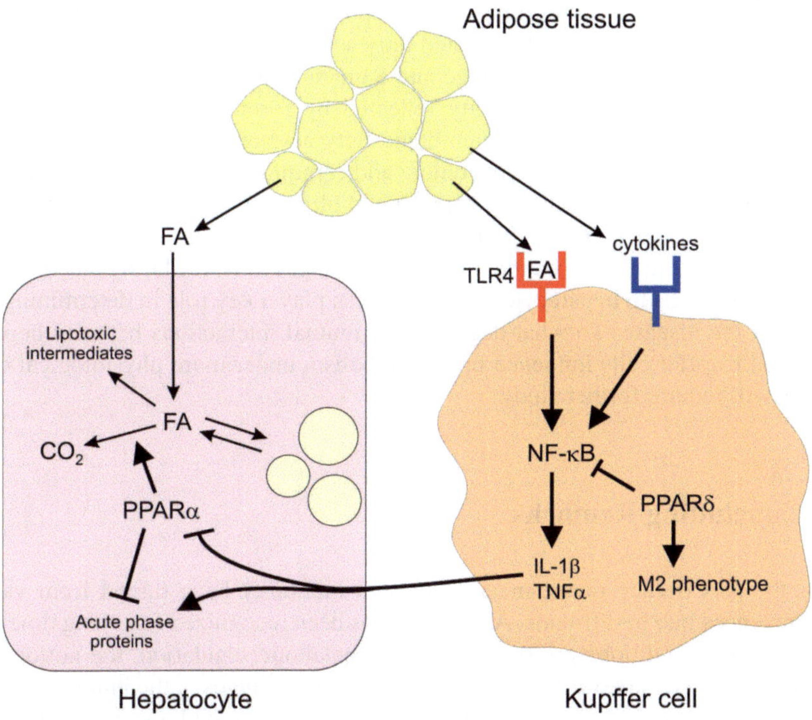


**Fig. 2** Model for interactions between hepatocytes and Kupffer cells influencing lipid metabolism and inflammatory responses. *FA* fatty acids; *TLR4* Toll-like receptor 4

steatosis or steatohepatitis was reduced, leading to improved insulin sensitivity (Huang et al. 2010; Neyrinck et al. 2009). Kupffer cells may stimulate triglyceride storage and suppress fatty acid oxidation in hepatocytes via TNFα (Huang et al. 2010). Alternatively, a role for Kupffer cell-derived IL-1β in promoting lipid accumulation via induction of Dgat2 expression or suppression of PPARα-dependent regulation of hepatic fatty acid oxidation has been proposed (Fig. 2) (Miura et al. 2010; Stienstra et al. 2010). The specific impact of Kupffer cells on lipid metabolism in hepatocytes may depend on their polarization state and inflammatory status. A recent study provided compelling evidence that the transcription factor PPARβ/δ may stimulate fatty acid oxidation and thereby protect against hepatic steatosis by promoting a M2-like phenotype in Kupffer cells (Odegaard et al. 2008).

Conversely, altered lipid metabolism and lipid accumulation in the hepatocyte may affect Kupffer cell activity (Baffy 2009). It is well established that hepatic steatosis increases the risk for inflammatory complications leading to nonalcoholic steatohepatitis. First, the increased space occupied by fatty hepatocytes may lead to impaired sinusoidal perfusion and entrapment of leukocytes, thereby engaging Kupffer cells. Second, impaired fatty acid uptake in hepatocytes and/or elevated lipolysis may increase exposure of Kupffer cells to fatty acids, leading to modulation of inflammatory pathways via interaction with cell surface receptors such as Toll-like receptor 4. Toll-like receptor 4 has been proposed as a relay system that mediates the proinflammatory effects of saturated fatty acids (Lee et al. 2001), although this concept has been challenged (Erridge and Samani 2009). Finally, excess lipids in hepatocytes may trigger lipotoxicity reflected by formation of reactive oxygen species, disturbances in cellular membrane fatty acid and phospholipid composition, and alterations of cholesterol content and ceramide signaling. These lipotoxic changes may cause hepatocellular damage and ultimately death, leading to activation of Kupffer cells (Trauner et al. 2010).

It is thus evident that throughout the various stages of NAFLD, signals traveling from Kupffer cells to hepatocytes and vice versa play a key role in determining the course of the disease. To what degree these mutual interactions between hepatocytes and Kupffer cells influence lipid metabolism under more physiological conditions will require further study.

# 4 Concluding Remarks

Over the past decade or so, metabolism and immunity have turned from vague acquaintances into best friends. Although it has been recognized for a long time that inflammatory conditions profoundly disturb metabolic regulation, the notion that metabolic diseases are often associated with marked changes in the immune system only gained acceptance more recently. The examples outlined above illustrate our rapidly growing insight into the molecular mechanisms that govern cross-talk between metabolic and immune-related signaling. These interconnections likely originated from the need of an organism to balance the energetic costs of immune responsiveness with its core mission of survival. It is expected that future research will further clarify specific mechanism of cross-talk between metabolic and immune regulation.

# References

Anderson EK, Gutierrez DA, Hasty AH (2010) Adipose tissue recruitment of leukocytes. Curr Opin Lipidol 21:172–177

Baffy G (2009) Kupffer cells in non-alcoholic fatty liver disease: the emerging view. J Hepatol 51:212–223

Beck IM, Vanden Berghe W, Vermeulen L, Yamamoto KR, Haegeman G, De Bosscher K (2009) Crosstalk in inflammation: the interplay of glucocorticoid receptor-based mechanisms and kinases and phosphatases. Endocr Rev 30:830–882

Bishop-Bailey D, Bystrom J (2009) Emerging roles of peroxisome proliferator-activated receptor-beta/delta in inflammation. Pharmacol Ther 124:141–150

Bougarne N, Paumelle R, Caron S, Hennuyer N, Mansouri R, Gervois P, Staels B, Haegeman G, De Bosscher K (2009) PPARalpha blocks glucocorticoid receptor alpha-mediated transactivation but cooperates with the activated glucocorticoid receptor alpha for transrepression on NF-kappaB. Proc Natl Acad Sci U S A 106:7397–7402

Boumpas DT, Chrousos GP, Wilder RL, Cupps TR, Balow JE (1993) Glucocorticoid therapy for immune-mediated diseases: basic and clinical correlates. Ann Intern Med 119:1198–1208
Clementi AH, Gaudy AM, van Rooijen N, Pierce RH, Mooney RA (2009) Loss of Kupffer cells in diet-induced obesity is associated with increased hepatic steatosis, STAT3 signaling, and further decreases in insulin signaling. Biochim Biophys Acta 1792:1062–1072
Commins SP, Watson PM, Padgett MA, Dudley A, Argyropoulos G, Gettys TW (1999) Induction of uncoupling protein expression in brown and white adipose tissue by leptin. Endocrinology 140:292–300
Cupps TR, Fauci AS (1982) Corticosteroid-mediated immunoregulation in man. Immunol Rev 65:133–155
Delerive P, Fruchart JC, Staels B (2001) Peroxisome proliferator-activated receptors in inflammation control. J Endocrinol 169:453–459
Erridge C, Samani NJ (2009) Saturated fatty acids do not directly stimulate Toll-like receptor signaling. Arterioscler Thromb Vasc Biol 29:1944–1949
Farooqi IS, O'Rahilly S (2009) Leptin: a pivotal regulator of human energy homeostasis. Am J Clin Nutr 89:980S–984S
Feuerer M, Herrero L, Cipolletta D, Naaz A, Wong J, Nayer A, Lee J, Goldfine AB, Benoist C, Shoelson S et al (2009) Lean, but not obese, fat is enriched for a unique population of regulatory T cells that affect metabolic parameters. Nat Med 15:930–939
Friedman JM, Halaas JL (1998) Leptin and the regulation of body weight in mammals. Nature 395:763–770
Gainsford T, Willson TA, Metcalf D, Handman E, McFarlane C, Ng A, Nicola NA, Alexander WS, Hilton DJ (1996) Leptin can induce proliferation, differentiation, and functional activation of hemopoietic cells. Proc Natl Acad Sci U S A 93:14564–14568
Gervois P, Kleemann R, Pilon A, Percevault F, Koenig W, Staels B, Kooistra T (2004) Global suppression of IL-6-induced acute phase response gene expression after chronic in vivo treatment with the peroxisome proliferator-activated receptor-alpha activator fenofibrate. J Biol Chem 279:16154–16160
Ghilardi N, Ziegler S, Wiestner A, Stoffel R, Heim MH, Skoda RC (1996) Defective STAT signaling by the leptin receptor in diabetic mice. Proc Natl Acad Sci U S A 93:6231–6235
Goldstein SA, Elwyn DH (1989) The effects of injury and sepsis on fuel utilization. Annu Rev Nutr 9:445–473
Hotamisligil GS, Shargill NS, Spiegelman BM (1993) Adipose expression of tumor necrosis factor-alpha: direct role in obesity-linked insulin resistance. Science 259:87–91
Howard JK, Lord GM, Matarese G, Vendetti S, Ghatei MA, Ritter MA, Lechler RI, Bloom SR (1999) Leptin protects mice from starvation-induced lymphoid atrophy and increases thymic cellularity in ob/ob mice. J Clin Invest 104:1051–1059
Huang W, Metlakunta A, Dedousis N, Zhang P, Sipula I, Dube JJ, Scott DK, O'Doherty RM (2010) Depletion of liver Kupffer cells prevents the development of diet-induced hepatic steatosis and insulin resistance. Diabetes 59:347–357
Imai E, Stromstedt PE, Quinn PG, Carlstedt-Duke J, Gustafsson JA, Granner DK (1990) Characterization of a complex glucocorticoid response unit in the phosphoenolpyruvate carboxykinase gene. Mol Cell Biol 10:4712–4719
Jump DB, Botolin D, Wang Y, Xu J, Christian B, Demeure O (2005) Fatty acid regulation of hepatic gene transcription. J Nutr 135:2503–2506
Kamei N, Tobe K, Suzuki R, Ohsugi M, Watanabe T, Kubota N, Ohtsuka-Kowatari N, Kumagai K, Sakamoto K, Kobayashi M et al (2006) Overexpression of monocyte chemoattractant protein-1 in adipose tissues causes macrophage recruitment and insulin resistance. J Biol Chem 281:26602–26614
Kanda H, Tateya S, Tamori Y, Kotani K, Hiasa K, Kitazawa R, Kitazawa S, Miyachi H, Maeda S, Egashira K et al (2006) MCP-1 contributes to macrophage infiltration into adipose tissue, insulin resistance, and hepatic steatosis in obesity. J Clin Invest 116:1494–1505

Kersten S, Desvergne B, Wahli W (2000) Roles of PPARs in health and disease. Nature 405:421–424

La Cava A, Matarese G (2004) The weight of leptin in immunity. Nat Rev Immunol 4:371–379

Lanthier N, Molendi-Coste O, Horsmans Y, van Rooijen N, Cani PD, Leclercq IA (2010) Kupffer cell activation is a causal factor for hepatic insulin resistance. Am J Physiol Gastrointest Liver Physiol 298:G107–116

Lee DE, Kehlenbrink S, Lee H, Hawkins M, Yudkin JS (2009) Getting the message across: mechanisms of physiological cross talk by adipose tissue. Am J Physiol Endocrinol Metab 296:E1210–1229

Lee JY, Sohn KH, Rhee SH, Hwang D (2001) Saturated fatty acids, but not unsaturated fatty acids, induce the expression of cyclooxygenase-2 mediated through Toll-like receptor 4. J Biol Chem 276:16683–16689

Lord GM, Matarese G, Howard JK, Baker RJ, Bloom SR, Lechler RI (1998) Leptin modulates the T-cell immune response and reverses starvation-induced immunosuppression. Nature 394:897–901

Lumeng CN, DelProposto JB, Westcott DJ, Saltiel AR (2008) Phenotypic switching of adipose tissue macrophages with obesity is generated by spatiotemporal differences in macrophage subtypes. Diabetes 57:3239–3246

Macfarlane DP, Forbes S, Walker BR (2008) Glucocorticoids and fatty acid metabolism in humans: fuelling fat redistribution in the metabolic syndrome. J Endocrinol 197:189–204

Mansouri RM, Bauge E, Staels B, Gervois P (2008) Systemic and distal repercussions of liver-specific peroxisome proliferator-activated receptor-alpha control of the acute-phase response. Endocrinology 149:3215–3223

Mattsson C, Olsson T (2007) Estrogens and glucocorticoid hormones in adipose tissue metabolism. Curr Med Chem 14:2918–2924

Minokoshi Y, Kim YB, Peroni OD, Fryer LG, Muller C, Carling D, Kahn BB (2002) Leptin stimulates fatty-acid oxidation by activating AMP-activated protein kinase. Nature 415:339–343

Miura K, Kodama Y, Inokuchi S, Schnabl B, Aoyama T, Ohnishi H, Olefsky JM, Brenner DA, and Seki E (2010) Toll-like receptor 9 promotes steatohepatitis by induction of interleukin-1beta in mice. Gastroenterology 139:323–334

Neyrinck AM, Cani PD, Dewulf EM, De Backer F, Bindels LB, Delzenne NM (2009) Critical role of Kupffer cells in the management of diet-induced diabetes and obesity. Biochem Biophys Res Commun 385:351–356

Nishimura S, Manabe I, Nagasaki M, Eto K, Yamashita H, Ohsugi M, Otsu M, Hara K, Ueki K, Sugiura S et al (2009) CD8+ effector T cells contribute to macrophage recruitment and adipose tissue inflammation in obesity. Nat Med 15:914–920

Odegaard JI, Ricardo-Gonzalez RR, Red Eagle A, Vats D, Morel CR, Goforth MH, Subramanian V, Mukundan L, Ferrante AW, Chawla A (2008) Alternative M2 activation of Kupffer cells by PPARdelta ameliorates obesity-induced insulin resistance. Cell Metab 7:496–507

Opherk C, Tronche F, Kellendonk C, Kohlmuller D, Schulze A, Schmid W, Schutz G (2004) Inactivation of the glucocorticoid receptor in hepatocytes leads to fasting hypoglycemia and ameliorates hyperglycemia in streptozotocin-induced diabetes mellitus. Mol Endocrinol 18:1346–1353

Pelleymounter MA, Cullen MJ, Baker MB, Hecht R, Winters D, Boone T, Collins F (1995) Effects of the obese gene product on body weight regulation in ob/ob mice. Science 269:540–543

Prentice AM, Hennig BJ, Fulford AJ (2008) Evolutionary origins of the obesity epidemic: natural selection of thrifty genes or genetic drift following predation release? Int J Obes (Lond) 32:1607–1610

Reichardt HM, Tuckermann JP, Gottlicher M, Vujic M, Weih F, Angel P, Herrlich P, Schutz G (2001) Repression of inflammatory responses in the absence of DNA binding by the glucocorticoid receptor. EMBO J 20:7168–7173

Rivera CA, Adegboyega P, van Rooijen N, Tagalicud A, Allman M, Wallace M (2007) Toll-like receptor-4 signaling and Kupffer cells play pivotal roles in the pathogenesis of non-alcoholic steatohepatitis. J Hepatol 47:571–579

Sanderson LM, de Groot PJ, Hooiveld GJ, Koppen A, Kalkhoven E, Muller M, Kersten S (2008) Effect of synthetic dietary triglycerides: a novel research paradigm for nutrigenomics. PLoS One 3:e1681

Shoelson SE, Herrero L, Naaz A (2007) Obesity, inflammation, and insulin resistance. Gastroenterology 132:2169–2180

Smoak KA, Cidlowski JA (2004) Mechanisms of glucocorticoid receptor signaling during inflammation. Mech Ageing Dev 125:697–706

Stienstra R, Mandard S, Patsouris D, Maass C, Kersten S, Muller M (2007) Peroxisome proliferator-activated receptor alpha protects against obesity-induced hepatic inflammation. Endocrinology 148:2753–2763

Stienstra R, Saudale F, Duval C, Keshtkar S, Groener JE, van Rooijen N, Staels B, Kersten S, Muller M (2010) Kupffer cells promote hepatic steatosis via interleukin-1beta-dependent suppression of peroxisome proliferator-activated receptor alpha activity. Hepatology 51:511–522

Straus DS, Glass CK (2007) Anti-inflammatory actions of PPAR ligands: new insights on cellular and molecular mechanisms. Trends Immunol 28:551–558

Strissel KJ, Defuria J, Shaul ME, Bennett G, Greenberg AS, Obin MS (2010) T-cell recruitment and Th1 polarization in adipose tissue during diet-induced obesity in C57BL/6 mice. Obesity (Silver Spring)

Tartaglia LA, Dembski M, Weng X, Deng N, Culpepper J, Devos R, Richards GJ, Campfield LA, Clark FT, Deeds J et al (1995) Identification and expression cloning of a leptin receptor OB-R. Cell 83:1263–1271

Trauner M, Arrese M, Wagner M (2010) Fatty liver and lipotoxicity. Biochim Biophys Acta 1801:299–310

Trayhurn P, Wood IS (2004) Adipokines: inflammation and the pleiotropic role of white adipose tissue. Br J Nutr 92:347–355

Valtonen TM, Kleino A, Rämet M, Rantala MJ (2010) Starvation reveals maintenance cost of humoral immunity. Evol Biol 37:49–57

van der Meer DL, Degenhardt T, Vaisanen S, de Groot PJ, Heinaniemi M, de Vries SC, Muller M, Carlberg C, Kersten S (2010) Profiling of promoter occupancy by PPAR{alpha} in human hepatoma cells via ChIP-chip analysis. Nucleic Acids Res 38:2839–2850

Weisberg SP, McCann D, Desai M, Rosenbaum M, Leibel RL, Ferrante AW Jr (2003) Obesity is associated with macrophage accumulation in adipose tissue. J Clin Invest 112:1796–1808

Winer S, Chan Y, Paltser G, Truong D, Tsui H, Bahrami J, Dorfman R, Wang Y, Zielenski J, Mastronardi F et al (2009) Normalization of obesity-associated insulin resistance through immunotherapy. Nat Med 15:921–929

Xu H, Barnes GT, Yang Q, Tan G, Yang D, Chou CJ, Sole J, Nichols A, Ross JS, Tartaglia LA et al (2003) Chronic inflammation in fat plays a crucial role in the development of obesity-related insulin resistance. J Clin Invest 112:1821–1830

Yu R, Kim CS, Kwon BS, Kawada T (2006) Mesenteric adipose tissue-derived monocyte chemoattractant protein-1 plays a crucial role in adipose tissue macrophage migration and activation in obese mice. Obesity (Silver Spring) 14:1353–1362

Zhang Y, Proenca R, Maffei M, Barone M, Leopold L, Friedman JM (1994) Positional cloning of the mouse obese gene and its human homologue. Nature 372:425–432

# Lipid Storage in Large and Small Rat Adipocytes by Vesicle-Associated Glycosylphosphatidylinositol-Anchored Proteins

**Günter Müller, Susanne Wied, Elisabeth-Ann Dearey, Eva-Maria Wetekam, and Gabriele Biemer-Daub**

**Abstract** Adipose tissue mass in mammals expands by increasing the average cell volume and/or total number of the adipocytes. Upregulated lipid storage in fully differentiated adipocytes resulting in their enlargement is well documented and thought to be a critical mechanism for the expansion of adipose tissue depots during the growth of both lean and obese animals and human beings. A novel molecular mechanism for the regulation of lipid storage and cell size in rat adipocytes was recently elucidated for the physiological stimuli, palmitate and $H_2O_2$, and the anti-diabetic sulfonylurea drug, glimepiride. It encompasses (1) the release of small vesicles, so-called adiposomes, harboring the glycosylphosphatidylinositol-anchored (c)AMP-degrading phosphodiesterase Gce1 and 5′-nucleotidase CD73 from donor adipocytes, (2) the transfer of the adiposomes and their interaction with detergent-insoluble glycolipid-enriched microdomains of the plasma membrane of acceptor adipocytes, (3) the translocation of Gce1 and CD73 from the adiposomes to the intracellular lipid droplets of the acceptor adipocytes, and (4) the degradation of (c) AMP at the lipid droplet surface zone by Gce1 and CD73 in the acceptor adipocytes, leading to the upregulation of the esterification of fatty acids into triacylglycerols and the downregulation of their release from triacylglycerols. This mechanism may provide novel strategies for the therapy of metabolic diseases, such as type 2 diabetes and obesity.

## 1 Introduction

It seems very likely that the size of adipocytes is regulated by multiple mechanisms, among them paracrine and endocrine feedback mechanisms (Raff 1996), which are currently understood as best for skeletal muscle. Here, growth and differentiation

G. Müller (✉), S. Wied, E.-A. Dearey, E.-M. Wetekam, and G. Biemer-Daub
Sanofi-Aventis Deutschland GmbH, Research & Development, 65926 Frankfurt, Germany
e-mail: Guenter.Mueller@sanofi-aventis.com

W. Meyerhof et al. (eds.), *Sensory and Metabolic Control of Energy Balance*, Results and Problems in Cell Differentiation 52,
DOI 10.1007/978-3-642-14426-4_3, 

factor 8 (GDF8), also known as myostatin, is secreted from myocytes and negatively regulates the generation of new muscle cells and thereby sets the size of the individual myocyte as well as that of the total muscle (Joulia-Ekaza and Cabello 2006). Loss-of-function mutations in GDF8 result in large increases in the size of the myocyte and muscle in animals and humans. The operation of similar feedback mechanisms that control adipocyte size is conceivable. It will be important to identify them at the molecular level since they may offer novel targets for pharmacological therapy when obesity is established after the final number of adipocytes in the body had been set.

## 2 Lipogenic and Antilipolytic Information Transfer Between Adipocytes

The following experimental evidence reported previously (Aoki et al. 2007; Müller et al. 2008a–e, 2009a, b, 2010a, b) suggests that there is communication between differently sized adipocytes about their esterifying and lipolytic states both in the absence and, considerably more pronounced, in the presence of certain stimuli via small vesicles, i.e., microvesicles and exosomes, harboring glycosylphosphatidylinositol-anchored proteins (GPI-proteins)(Nosjean et al. 1997), so-called adiposomes (ADIP)(Aoki et al. 2007): (1) Older and larger adipocytes are more efficient in releasing the GPI-proteins, Gce1 and CD73, incorporated into ADIP than younger and smaller ones and thus operate as the "better" donors. (2) Upon incubation with those ADIP younger and smaller adipocytes are more efficient in translocating Gce1 and CD73 from the ADIP to cytoplasmic lipid droplets (LD) than older and larger adipocytes and thus operate as the "better" acceptors. (3) Mixed populations of about equal numbers of large and small adipocytes are more efficient in esterification stimulation and lipolysis inhibition in the basal state (Fig. 1) as well as in response to palmitate, the antidiabetic drug glimepiride, and $H_2O_2$ (Fig. 2) than separate populations of the same total number of large and small adipocytes (calculated as the arithmetic means) with regard to both maximal responsiveness (i.e., increase in fold-stimulation and %-inhibition at maximally effective concentrations) and sensitivity (i.e., decrease in $EC_{50}$-values). (4) The responsiveness of mixed populations of about equal numbers of large and small adipocytes for esterification stimulation and lipolysis inhibition by palmitate, glimepiride, and $H_2O_2$ becomes reduced upon removal of the Gce1- and CD73-harboring ADIP, which in contrast does not affect populations of either large or small adipocytes.

The apparent dependence of the basal as well as palmitate-, glimepiride-, and $H_2O_2$-regulated lipid metabolism on the presence of differently sized adipocytes and of Gce1- and CD73-harboring ADIP in the incubation mixture is compatible with a hypothetical model (Fig. 3) for transferring ADIP-encoded information about the relatively high esterifying and lipolytic states from older and larger adipocytes (a) to

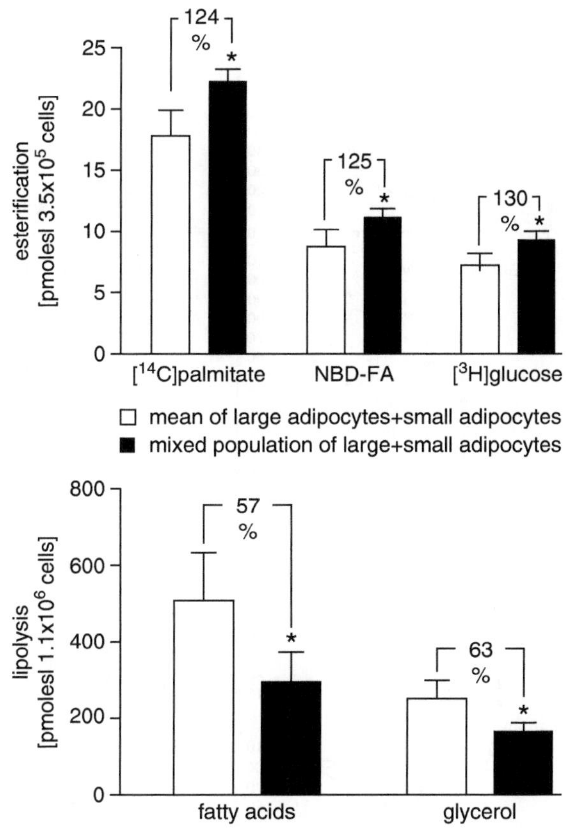


**Fig. 1** Basal esterification and lipolysis in separate vs. mixed populations of large and small adipocytes. Separate or mixed (2:1) populations of large and small adipocytes were assayed for esterification (3 h, 37°C) or isoproterenol (0.1 μM)-induced lipolysis (2 h, 37°C). Esterification and lipolysis were calculated as the amounts of [U-$^{14}$C]palmitate, fluorescently labelled fatty acid (NBD-FA) or [3-$^{3}$H]glucose incorporated into triacylglycerol and the amounts of fatty acids or glycerol released into the incubation medium, respectively, by identical cell numbers of the separate and mixed populations, each. The arithmetic means calculated for the separate small and large adipocytes (*open bars*) and the values measured for the mixed populations of small and large adipocytes (*filled bars*) are given for identical total cell numbers. Means ± SD were derived from 3 to 5 incubations, each, with determinations in triplicate with significant differences of the mixed vs. the separate populations indicated by $*p \leq 0.05$. In addition, esterification and lipolysis rates of the mixed populations are given as percentages of the arithmetic means calculated for the separate populations

younger and smaller adipocytes, which exhibit relatively low esterification and lipolysis (c). This ADIP-based information transfer operates even in the basal state but is considerably upregulated by certain exogenous esterifying and antilipolytic stimuli, such as palmitate, glimepiride, or $H_2O_2$ (b), but not insulin. Following information transfer from the large donor (b) to the small acceptor (d) adipocytes, the latter acquire the potential for more pronounced esterification and less pronounced lipolysis. This is based on the upregulated degradation of (c)AMP at

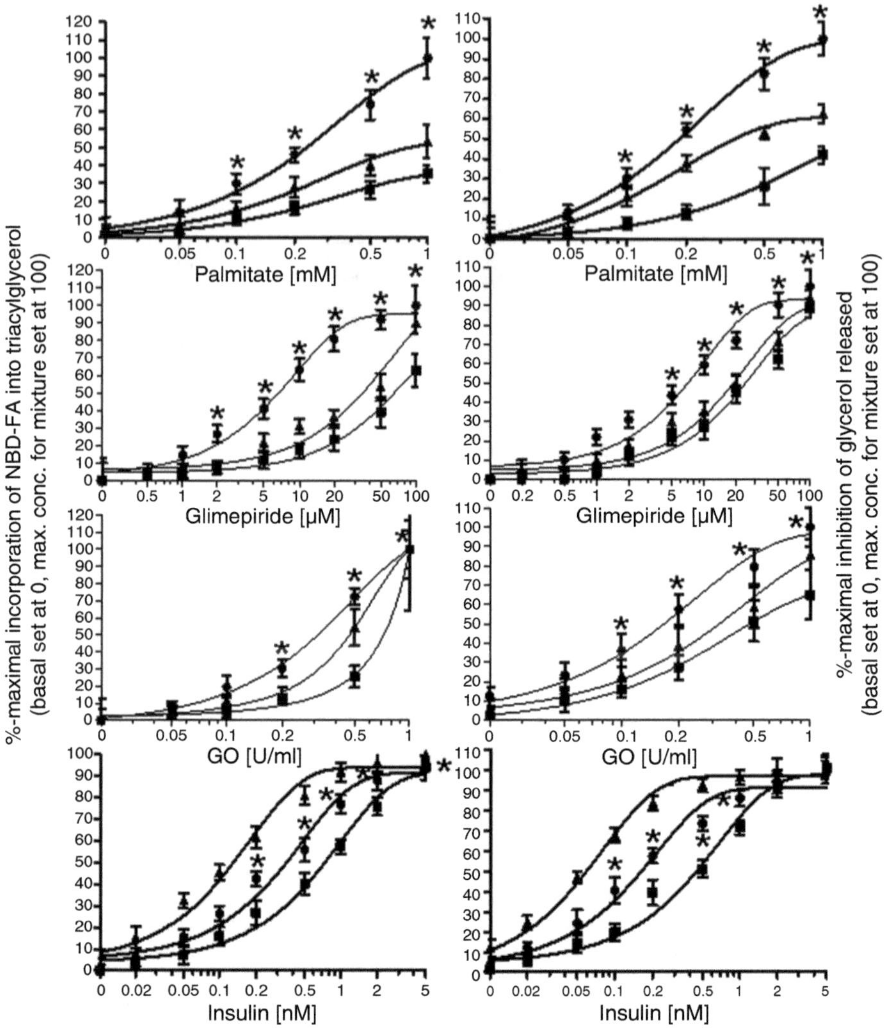


**Fig. 2** Esterification stimulation and lipolysis inhibition by palmitate, glimepiride and $H_2O_2$ in separate vs. mixed populations of large and small adipocytes. After incubation (30 min, 37°C) of separate or mixed (2:1) populations of large and small adipocytes in the absence (basal state) or presence of increasing concentrations of palmitate, glimepiride, glucose oxidase (GO; for the generation of $H_2O_2$ in the incubation medium) or insulin. The adipocytes were recovered by flotation (500 g, 2 min, 20°C) and then assayed for esterification (3 h, 37°C) or isoproterenol (0.1 μM isoproterenol)-induced lipolysis (2 h, 37°C). Esterification stimulation and lipolysis inhibition in response to each stimulus are given as percentages of the amounts of NBD-FA incorporated into triacylglycerol (*left panels*) and of the reductions in the amounts of glycerol released into the incubation medium (*right panels*) by the separate small (*triangle*) or large (*square*) adipocyte populations or the mixed populations (*circle*) with basal set at 0 and the maximally effective concentration of this stimulus in the mixed populations set at 100. Mean ± SD were derived from 2 to 3 adipocyte incubations, each, with determinations in duplicate with significant differences of the mixed populations vs. both separate small and large adipocyte populations indicated by $*p \leq 0.05$

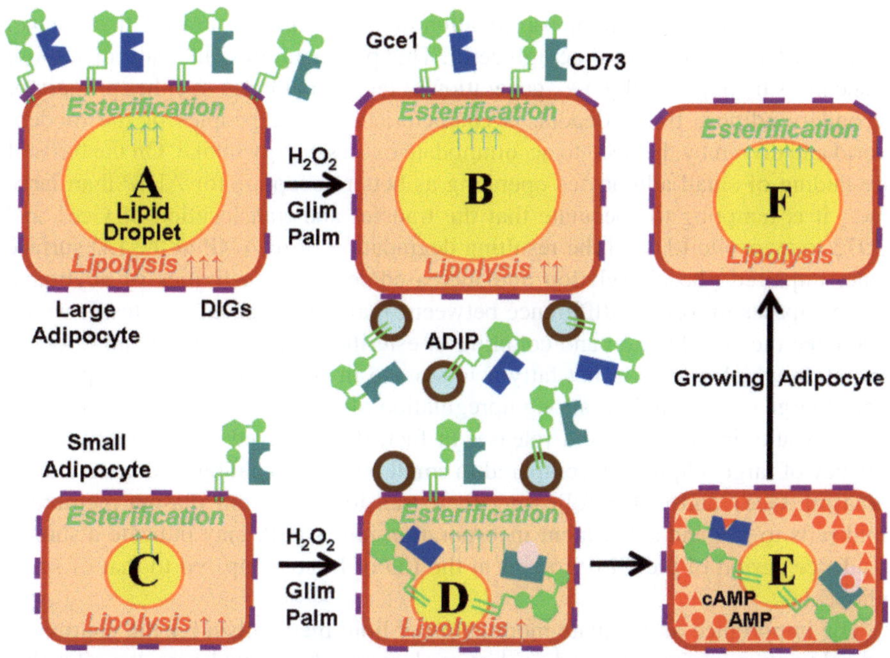


**Fig. 3** Hypothetical model for the paracrine regulation of lipid storage within adipose tissue depots by palmitate, glimepiride and $H_2O_2$. (**a** and **c**) In the basal state Gce1 and CD73 are predominantly located at plasma membrane lipid rafts of large and small adipocytes, which display basal esterification and lipolysis at medium and low rates, respectively. (**b**) Upon exposure of large (donor) adipocytes to palmitate (Palm), glimepiride (Glim) and $H_2O_2$, esterification becomes upregulated and lipolysis downregulated to moderate degrees, each, with Gce1/CD73 being incorporated into the phospholipid bilayer of ADIP and then released into the interstitial space. (**d**) Upon exposure of small (acceptor) adipocytes to palmitate, glimepiride and $H_2O_2$, esterification becomes upregulated and lipolysis downregulated to a high degree. This is accompanied by the release of ADIP-associated Gce1 and CD73 into the interstitial space which subsequently interact with plasma membrane lipid rafts and finally translocate to cytoplasmic LD of the small adipocytes. (**e**) Upon anchorage of Gce1/CD73 at the LD phospholipid monolayer in small adipocytes cAMP/AMP are degraded leading to decreased cAMP/AMP concentrations at the LD surface zone. (**f**) The resulting diminished cAMP-dependent phosphorylation of key enzymes of lipid metabolism, such as glycerol-3-phosphate acyltransferase and hormone-sensitive lipase, located at the LD surface zone triggers the coordinated upregulation of esterification and downregulation of lipolysis, respectively, to a very high degree, each. The consequent stimulation of LD biogenesis is paralleled by increase in size of the (initially small) adipocytes

the LD surface zone by (c)AMP-degrading phosphodiesterase Gce1 and 5′-nucleotidase CD73 (e). The fostered accumulation of triacylglycerol in concert with enhanced LD biogenesis ultimately leads to size enlargement of basal and, more pronouncedly, stimulus-induced adipocytes (f).

The increase in cell size in course of coordinated esterification (stimulation) and lipolysis (inhibition) via the transfer of ADIP-associated Gce1 and CD73 to young and small adipocytes is thus compatible with their physiological function, i.e., the net storage of fatty acids upon esterification into triacylglycerol rather than the net release of fatty acids from triacylglycerol. This postmaturation increase in adipose tissue mass by hypertrophy, i.e., generation of large adipocytes in either primary or secondary adipose tissue depots, is characteristic of the peoples of the modern world and driven by the need to accommodate excess energy intake. On the basis of the finding of small adipocytes operating as better acceptors for ADIP than large ones, it is tempting to speculate that the transfer and translocation of Gce1 and CD73 to cytosolic LD and the resulting degradation of (c)AMP at the LD surface zone, impaired phosphorylation and increased/decreased activities of acyltransferases/lipases makes the difference between small and large adipocytes. It would guarantee the coordination and coupling of esterification stimulation and lipolysis inhibition which results in net fatty acid storage in small adipocytes compared to that of large ones with their strong upregulation of both esterification and lipolysis which results in net fatty acid release. In fact, the observed considerably higher efficacy of large adipocytes compared to small ones in operating as donors, i.e., in releasing Gce1 and CD73 following their translocation from plasma membrane lipid rafts to LD and subsequent incorporation into ADIP, may indicate a shorter residence period of the GPI-proteins at the LD of large compared to that of small adipocytes.

The importance of adequate modulation of lipid metabolism for the pharmacological treatment of metabolic disorders, such as metabolic syndrome, type 2 diabetes, and obesity, has been fully recognized during the last decade (Moller 2001; Shi and Burn 2004; Langin 2006, 2010; Pilch and Bergenheim 2006; Pfeiffer 2007; Ahmadian et al. 2009; Fernandez-Veledo et al. 2009; Chavez and Summers 2010). The putative paracrine and/or endocrine transfer of lipogenic and antilipolytic information via ADIP-associated GPI-proteins may shift the burden of lipid loading and storage from large to small adipocytes within the same or distinct adipose tissue depots (Fig. 3). It is also conceivable that ADIP manage to leave the adipose tissue depots and be distributed via the circulation as is true for microvesicles and exosomes released from blood, endothelial and solid tumor cells (Freyssinet 2003; Fevrier and Raposo 2004; Janowska-Wieczorek et al. 2005; Koga et al. 2005). Thereby, ADIP may reach insulin target tissues, such as muscle and liver, and even insulin-producing pancreatic β-cells. In case of operation of those cells as acceptor cells for the ADIP-associated GPI-proteins as has been demonstrated so far for adipocytes, Gce1 and CD73 would be transferred to the plasma membrane lipid rafts and then translocated to the existing, albeit small, LD of muscle, liver and β-cells (Björntorp and Karlsson 1970), where they degrade cAMP at the LD surface zone. The resulting upregulation of esterification and downregulation of lipolysis would ultimately lead to growth of the LD in these nonadipose tissue cells and impair their insulin responsiveness and sensitivity. The underlying lipotoxic mechanisms are ill-defined at the molecular level, but seem to encompass intermediates of the synthesis and degradation of triacylglycerol rather than triacylglycerol molecules themselves. Thus, the putative transfer of antilipolytic and lipogenic information between adipose tissues and from adipose tissues to metabolically relevant nonadipose tissues could per se or in combination contribute to the development of obesity-driven insulin resistance and thus may provide novel targets for the pharmacological therapy of metabolic diseases. Moreover, a multitude of novel roles for LD in adipose and nonadipose cells is currently being elucidated. These encompass, but apparently are not restricted to, the storage and scavenging of lipophilic and toxic substances as well as to the sequestration, chaperoning and release for translocation to other compartments, such as the nucleus (e.g., histones) or plasma membrane lipid rafts and adiposomes (e.g. Gce1, CD73) of a variety of regulatory, signaling and metabolic proteins. This strongly argues for the usefulness of targeting LD for the therapy of a number of diseases by the transfer and translocation of appropriate GPI-proteins from in vitro reconstituted ADIP upon their administration to the patient. The feasibility of this exciting putative mode for the controlled expression of (recombinant and modified) proteins at the cytoplasmic face of LD in diseased adipose and nonadipose cells without the use of gene therapy has to be addressed in future preclinical and clinical studies.

## 3 Conclusion

Taken together, the recent findings argue for the transfer of information coding for esterification stimulation and lipolysis inhibition from large to small adipocytes via microvesicles and/or exosomes harboring glycosylphosphatidylinositol-anchored proteins, so-called adiposomes. Since this transfer ultimately leads to the formation of lipid droplets and the maturation of small to large adipocytes, its tissue-specific modulation may represent a novel target for the pharmacological therapy of metabolic diseases, such as obesity and type 2 diabetes.

## References

Ahmadian M, Duncan RE, Sul HS (2009) The skinny on fat: lipolysis and fatty acid utilization in adipocytes. Trends Endocrinol Metab 20:424–428

Aoki N, Jin-no S, Nakagawa Y, Asai N, Arakawa E, Tamura N, Tamura T, Matsuda T (2007) Identification and characterization of microvesicles secreted by 3 T3-L1 adipocytes: redox- and hormone-dependent induction of milk fat globule-epidermal growth factor 8-associated microvesicles. Endocrinology 148:3850–3862

Björntorp P, Karlsson M (1970) Triglyceride synthesis in human subcutaneous adipose tissue cells of different size. Eur J Clin Invest 1:112–117

Chavez JA, Summers SA (2010) Lipid oversupply, selective insulin resistance, and lipotoxicity: molecular mechanisms. Biochim Biophys Acta 1801:252–265

Fernandez-Veledo S, Nieto-Vazquez I, Vila-Bedmar R, Garcia-Guerra L, Alonso-Chamorro M, Lorenzo M (2009) Molecular mechanisms involved in obesity-associated insulin resistance: therapeutical approach. Arch Physiol Biochem 115:227–239

Fevrier B, Raposo G (2004) Exosomes: endosomal-derived vesicles shipping extracellular messages. Curr Opin Cell Biol 16:415–421

Freyssinet J-M (2003) Cellular microparticles: what are they bad or good for? J Thromb Haemost 1:1655–1662
Janowska-Wieczorek A, Wysoczynski M, Kijowski J, Marquez-Curtis L, Machalinski B, Ratajczak J, Ratajczak MZ (2005) Microvesicles derived from activated platelets induce metastasis and angiogenesis in lung cancer. Int J Cancer 113:752–760
Joulia-Ekaza D, Cabello G (2006) Myostatin regulation of muscle development: molecular basis, natural mutations, physiopathological aspects. Exp Cell Res 312:2401–2414
Koga K, Matsumoto K, Akiyoshi T, Kubo M, Yamanaka N, Tasaki A, Nakashima H, Nakamura M, Kuroki S, Tanaka M, Katano M (2005) Purification, characterization and biological significance of tumor-derived exosomes. Anticancer Res 25:3703–3707
Langin D (2006) Adipose tissue lipolysis as a metabolic pathway to define pharmacological strategies against obesity and the metabolic syndrome. Pharmacol Res 53:482–491
Langin D (2010) Recruitment of brown fat and conversion of white into brown adipocytes: strategies to fight the metabolic complications of obesity. Biochim Biophys Acta 1801:372–376
Moller DE (2001) New drug targets for type 2 diabetes and the metabolic syndrome. Nature 414:821–827
Müller G, Jung C, Straub J, Wied S (2009a) Induced release of membrane vesicles and exosomes from rat adipocytes containing lipid droplet, lipid raft and glycosylphosphatidylinositol-anchored proteins. Cell Signal 21:324–338
Müller G, Jung C, Wied S, Biemer-Daub G (2009b) Induced translocation of glycosylphosphatidylinositol-anchored proteins from lipid droplets to adiposomes in rat adipocytes. Br J Pharmacol 158:749–770
Müller G, Wied S, Jung C, Biemer-Daub G, Frick W (2010a) Transfer of glycosylphosphatidylinositol-anchored 5′-nucleotidase CD73 from adiposomes into rat adipocytes stimulates lipid synthesis. Br J Pharmacol. Br J Pharmacol 160:878–891
Müller G, Wied S, Jung C, Frick W, Biemer-Daub G (2010b) Inhibition of lipolysis by adiposomes containing glycosylphosphatidylinositol-anchored Gce1 protein in rat adipocytes. Arch Physiol Biochem 116:28–41
Müller G, Over S, Wied S, Frick W (2008a) Association of (c)AMP-degrading glycosylphosphatidylinositol-anchored proteins with lipid droplets is induced by palmitate, $H_2O_2$ and the sulfonylurea drug, glimepiride, in rat adipocytes. Biochemistry 47:1274–1287
Müller G, Wied S, Jung C, Over S (2008b) Translocation of glycosylphosphatidylinositol-anchored proteins to lipid droplets and inhibition of lipolysis in rat adipocytes is mediated by reactive oxygen species. Br J Pharmacol 154:901–913
Müller G, Wied S, Jung C, Straub J (2008c) Coordinated regulation of esterification and lipolysis by palmitate, $H_2O_2$ and the anti-diabetic sulfonylurea drug, glimepiride, in rat adipocytes. Eur J Pharmacol 597:6–18
Müller G, Wied S, Over S, Frick W (2008d) Inhibition of lipolysis by palmitate, $H_2O_2$ and the sulfonylurea drug, glimepiride, in rat adipocytes depends on cAMP degradation by lipid droplets. Biochemistry 47:1259–1273
Müller G, Wied S, Walz N, Jung C (2008e) Translocation of glycosylphosphatidylinositol-anchored proteins from plasma membrane microdomains to lipid droplets in rat adipocytes is induced by palmitate, $H_2O_2$ and the sulfonylurea drug, glimepiride. Mol Pharmacol 73:1513–1529
Nosjean O, Briolay A, Roux B (1997) Mammalian GPI proteins: sorting, membrane residence and functions. Biochim Biophys Acta 1331:153–186
Pfeiffer AFH (2007) Adipose tissue and diabetes therapy: do we hit the target? Horm Metab Res 39:734–738
Pilch PF, Bergenheim N (2006) Pharmacological targeting of adipocytes/fat metabolism for treatment of obesity and diabetes. Mol Pharmacol 70:779–785
Raff MC (1996) Size control: the regulation of cell numbers in animal development. Cell 26:173–175
Shi Y, Burn P (2004) Lipid metabolizing enzymes: emerging drug targets for the treatment of obesity. Nat Rev Drug Discov 3:695–710

# Autophagy and Regulation of Lipid Metabolism

**Rajat Singh**

**Abstract** Macroautophagy (henceforth referred to as autophagy) is an in-bulk lysosomal degradative pathway that plays a crucial role in the maintenance of cellular homeostasis through the removal of damaged proteins and aged organelles. Following nutrient deprivation, a primary cellular response is the induction of autophagy that breaks down redundant cellular components and provides amino acids and additional precursor molecules for processes critical for cellular survival. In parallel, nutrient depletion leads to the mobilization of cellular lipid stores to supply free fatty acids for energy, thus pointing to regulatory and functional similarities between autophagy and lipid metabolism. The current chapter discusses the novel and mutually exclusive roles of autophagy in the regulation of lipid metabolism in the liver and of fat storage within the adipose tissue. Our studies in cultured hepatocytes and the murine liver have demonstrated that autophagy serves to degrade intracellular lipid stores through a process that we have termed "macrolipophagy" and that ablation of liver-specific autophagy leads to excessive hepatic lipid accumulation and the development of fatty liver. In contrast, preadipocytes in culture that lacked autophagy failed to differentiate into mature adipocytes and exhibited a reduction in fat storage that translated to decreased adipose tissue mass in an *in vivo* mouse model. These recent findings establish an association between autophagy and regulation of hepatic lipid metabolism and adipose tissue biology, thus providing new mechanistic insights into the regulation of these complex processes. These findings also highlight the possibility of novel therapeutic approaches, such as differential organ-specific regulation of autophagy to solve problems that arise from lipid over accumulation that occur in the metabolic syndrome and with aging.

R. Singh
Department of Medicine (Endocrinology) and Molecular Pharmacology, Albert Einstein College of Medicine, Forchheimer 505, 1300 Morris Park Avenue, Bronx, NY 10461, USA
e-mail: rajat.singh@einstein.yu.edu

W. Meyerhof et al. (eds.), *Sensory and Metabolic Control of Energy Balance*, Results and Problems in Cell Differentiation 52,
DOI 10.1007/978-3-642-14426-4_4, 

# 1 An Introduction to Autophagy

Autophagy or "self-eating" is a conserved cellular process critical for maintaining cellular homeostasis (Cuervo 2008) through the targeting of altered cytosolic proteins, aged organelles and redundant cytosol, and even pathogenic organisms to the lysosomes for degradation. Besides this crucial role in quality control, autophagy has also been shown to be involved in a host of other functions such as growth and differentiation (Mizushima 2009), metabolic regulation (Singh et al. 2009a, b) and as an alternative energy source following nutrient deficiency. Three different forms of autophagy have been described in mammalian systems (Fig. 1), *macroautophagy*, *microautophagy*, and *chaperone-mediated autophagy* (CMA) (Esclatine et al. 2009; Kon and Cuervo 2010). These three forms of autophagy co-exist in most cells at a given moment, although the induction of these discrete forms of autophagy differs temporally. For instance, under baseline cellular conditions all forms of autophagy functions at a basal level, however, following nutrient deficiency, macroautophagy is induced first and acquires maximal activation at 6–8 h of continuous stress (Esclatine et al. 2009). If the given stress persists beyond 6–8 h, macroautophagy gradually tapers off and induction of CMA occurs that

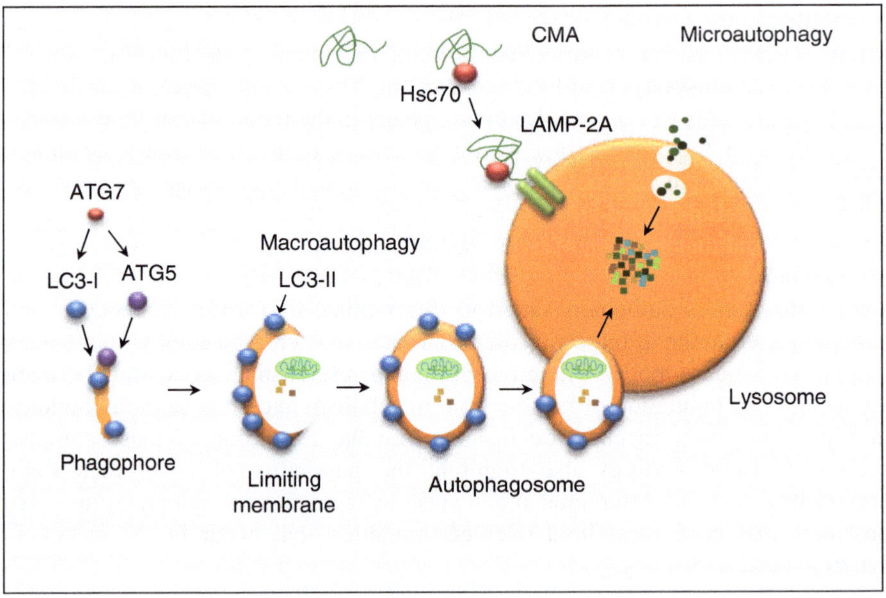


**Fig. 1** Types of autophagy in mammalian cells. Three types of autophagy have been described in mammalian cells, *macroautophagy*, an in-bulk degradative pathway that requires the de novo synthesis of a limiting membrane for sequestration and delivery of cargo to the lysosome. *Microautophagy* involves engulfment of cytosolic material by the lysosomal membrane itself. *Chaperone-mediated autophagy* targets specific proteins containing the recognition template, *KFERQ motif* to lysosomal membrane-associated translocation complex, LAMP-2A, for substrate internalization and degradation

attains maximal activation by 12–24 h (Kon and Cuervo 2010) and is maintained until the stress dissipates. Autophagy can also be qualitatively classified based upon the type of cargo that is being degraded, for instance, mitophagy mediates mitochondrial degradation (Tolkovsky 2009). Likewise, ribophagy (Beau et al. 2008), reticulophagy (Tasdemir et al. 2007), and pexophagy (Manjithaya et al. 2010) are required for the degradation of ribosomes, endoplasmic reticulum, and peroxisomes, respectively. Very recently, macroautophagy has also been shown to mediate the degradation of cellular lipid droplets under basal conditions as well as when macroautophagy is induced following lipogenic stimuli, by a process termed macrolipophagy (Singh et al. 2009a). This function of macroautophagy in the regulation of cellular lipid stores will be the focus of part of this chapter.

## *1.1 Types of Autophagy*

*Macroautophagy* is an in-bulk lysosomal degradative pathway that involves the initial de novo formation of a phagophore (He and Klionsky 2009), which then adds on additional membranes through yet unclear mechanisms and forms a limiting membrane. The limiting membrane sequesters portions of the cytosol destined for degradation and eventually seals upon itself to form a double-walled autophagosome (He and Klionsky 2009). The autophagosome lacks the acidic environment and the enzymes required for terminal digestion of the engulfed contents; thus, content degradation occurs via the fusion of the autophagosome with the lysosome that supplies the acidic environment as well as a battery of hydrolases. *Microautophagy*, a second form of autophagy, involves engulfment of cargo by the lysosomal membrane itself (Esclatine et al. 2009). Single-membrane vesicles that contain the cargo are pinched off the lysosomal membrane and rapidly degraded in the lysosomal lumen. A third form of autophagy, *CMA*, is highly specific for an estimate of 30% of the soluble cytosolic proteins (Kon and Cuervo 2010). In *CMA*, proteins that contain a recognition template, the KFERQ motif, are recognized by the cytosolic chaperone hsc70 and directed to the lysosome (Kon and Cuervo 2010). Upon the lysosomal surface, the protein–hsc70 complex is recognized by the lysosome-associated membrane protein (LAMP-2A) membrane receptor. The protein is then unfolded and internalized into the lysosome through the translocation complex comprised mainly of the LAMP-2A receptor (Cuervo and Dice 1996). The current chapter focuses on the role of macroautophagy in lipid metabolism; thus, the two additional forms of autophagy, microautophagy, and CMA will not be discussed any further.

## *1.2 Regulation of Macroautophagy*

The induction of macroautophagy (henceforth termed autophagy) occurs following conditions such as cellular stress or starvation leading to the formation of autophagosomes. Autophagosome formation is a complex and highly regulated process that requires more than 30 autophagy-related proteins (ATG) that were identified by molecular dissection of the autophagic process through yeast genetic screens (He and Klionsky 2009). These ATG proteins form functional complexes that mediate individual steps of autophagy; initiation or induction, nucleation, membrane elongation, cargo recognition, and the fusion of autophagosomes with lysosomes.

The *induction* of autophagy requires the activation of the autophagy initiation complex or the class III phosphatidylinositol-3-kinase (PI3K) (Fig. 2a), which occurs through the release of Beclin-1 (ATG6 in the yeast) from the Bcl-2–Beclin-1 complex through starvation or stress-induced phosphorylation of Bcl-2 (Wei et al. 2008). Beclin-1 is then recruited along with additional molecules, vps15, vps30, vps34, ATG14L, and UV radiation-resistance associated gene protein (UVRAG) to form the active class III PI3K complex (Petiot et al. 2000; Tassa et al. 2003). The activation of class III PI3K allows the mobilization of this induction complex to the site of autophagosome formation. A process called *nucleation* involves the mobilization of this initiation complex to the site of the limiting membrane formation (He and Klionsky 2009). In addition, lipid phosphorylation by the PI3K complex is critical for the recruitment of additional ATG molecules to the phagophore leading to elongation of the limiting membrane. The formation of the limiting membrane requires two parallel conjugation cascades, the microtubule-associated protein 1 light chain 3 (LC3) or ATG8 and the ATG5–12 conjugation cascades (Fig. 2b) that are similar to the ubiquitin conjugation system involving discrete ligases for activation and enzymatic conjugation of substrates (He and Klionsky 2009). These conjugation events occur on the surface of the limiting membrane and are crucial for membrane elongation. The ubiquitin-like ligase ATG7 is required for conjugation of ATG5 and ATG12 to form the ATG5–12 complex. Independently, cleavage and activation of cytosolic LC3-I occurs by the proteolytic cleavage of a cysteinyl residue by the protease ATG4. Activated LC3-I acquires a phosphatidylethanolamine residue that then forms the membrane-associated LC3-II (Kabeya et al. 2000) through a reaction that requires ATG7, in addition to the E2-like activity of ATG3. The molecular mechanisms of the later steps that mediate sealing of the limiting membrane to form autophagosomes and fusion events that form autophagolysosomes are yet unclear. However, it is thought that additional ATGs, specific SNARE, and Rab proteins as well as cytoskeletal elements may be involved in these fusion events (He and Klionsky 2009).

Autophagy is centrally regulated by the upstream serine/threonine kinase mTOR (mammalian target of rapamycin) (Jung et al. 2009). Under conditions of nutrient deprivation or following rapamycin treatment, inhibition of mTOR occurs leading to activation of Unc-51-like kinase-1 and -2 (ULK) that phosphorylates and activates ATG13 and FIP200 (focal adhesion kinase family-interacting protein of 200 kDa), thus, leading to the induction of autophagy (Jung et al. 2009). Nutritional excess, in particular, amino acids activate mTOR that phosphorylates and inactivates the ULK–ATG13–FIP200 complex resulting in the switching off of autophagy (Jung et al. 2009). Another crucial regulator of starvation-induced

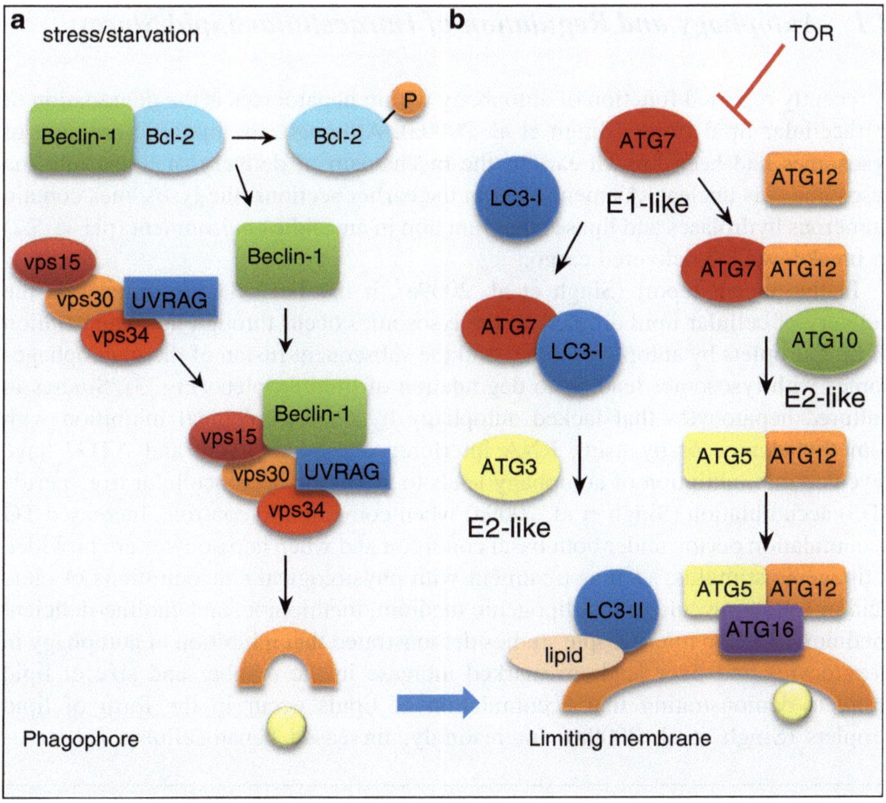


**Fig. 2** Autophagy induction complex (**a**) and (**b**) conjugation cascades that lead to autophagosome formation. The induction of autophagy requires release of Beclin-1 from Bcl-2 and its binding to vps15, vps30, vps34, and UVRAG to form the active Class III PI3K complex. Lipid phosphorylation by active class III PI3K recruits the two conjugation cascades to the site of autophagosome formation. The LC3 and ATG5–12 conjugation cascades require the ubiquitin-like ligase activity of ATG7. Products of these two cascades, LC3-II and the ATG5–12–16 conjugate, form structural components of the limiting membrane

autophagy is c-Jun *N*-terminal kinase that mediates the phosphorylation of Bcl-2 to release Beclin-1 in response to nutrient deprivation (Wei et al. 2008).

## 2 Autophagy and Lipid Metabolism

An established role of autophagy is the degradation of cellular organelles, redundant cytosol and proteins. However, only in these recent years, some reports have demonstrated a novel relationship between autophagy and cellular lipid metabolism, thus unfolding a new area of research.

## 2.1 Autophagy and Regulation of Intracellular Lipid Stores

A recently reported function of autophagy within hepatocytes is the degradation of intracellular lipid stores (Singh et al. 2009a). Although, the lipolytic function of lysosomes had been known earlier, the mechanism of delivery of lipids into the lysosome was unclear. As mentioned in the earlier sections, the lysosomes contain numerous hydrolases and lipases that function in an acidic environment (pH $< 5.2$) to breakdown the delivered cargo.

In this recent report (Singh et al. 2009a), it has been demonstrated that the delivery of cellular lipid droplets to the lysosomes occur through the sequestration of lipid droplets by autophagosomes and the subsequent fusion of these autophagosomes with lysosomes leading to degradation of lipid droplets (Fig. 3). Studies in cultured hepatocytes that lacked autophagy by pharmacological inhibition with 3-methyladenine or by using RNA interference against ATG5 and ATG7 have revealed that inhibition of autophagy leads to increased hepatocellular triglyceride (TG) accumulation (Singh et al. 2009a) when compared to controls. Increased TG accumulation occurs under both basal condition and when hepatocytes are provided a lipogenic stimulus, such as treatment with physiological concentrations of oleic acid or following culture in a lipogenic medium, methionine, and choline-deficient medium. Electron microscopic studies demonstrated that inhibition of autophagy in hepatocytes and liver leads to marked increase in the number and size of lipid droplets demonstrating that accumulation of lipids occur in the form of lipid droplets (Singh et al. 2009a). Interestingly, increased hepatocellular TG stores

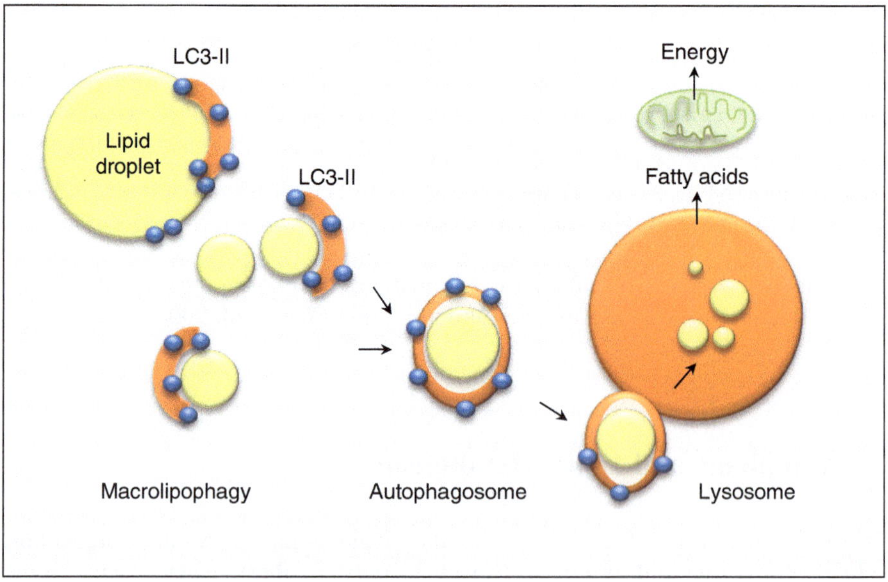


**Fig. 3** The process of macrolipophagy occurs by the de novo formation of a limiting membrane that sequesters cytosolic lipid droplets and delivers the lipid cargo to the lysosomes for degradation

results from decreased lipolysis of lipid stores due to decreased delivery of lipid cargo into the lysosomes, and not from increased hepatocellular TG synthesis or a reduction in secretion in the form of VLDL. Immunofluorescence co-localization experiments between a neutral lipid dye (bodipy 493/503) and the autophagosomal marker (LC3) or the lysosomal marker (LAMP1) have revealed co-localization of cellular lipids with components of the autophagosomal and lysosomal system under conditions that activate autophagy (Singh et al. 2009a), such as treatment with rapamycin (inhibitor of TOR) or provision of a lipid stimulus. Furthermore, pharmacological or genetic ablation of autophagy decreased the observed bodipy-LC3 and bodipy-LAMP1 co-localizations.

In mammals, induction of autophagy occurs in response to starvation and consequently, lipid droplet components such as structural proteins TIP47 and adipophilin (ADRP) could be detected within isolated autophagic vacuoles and lysosomes from starved mice liver (Singh et al. 2009a). Conversely, immunoblotting demonstrated increased autophagosome-associated LC3-II levels within lipid droplets fractions isolated from livers of starved rodents (Singh et al. 2009a). Additionally, immunoelectronmicrographs demonstrated increased presence of LC3-positive membranes on lipid droplets in liver sections from starved animals. Thus, these *in vivo* studies have demonstrated a functional interaction between cellular lipid stores and the autophagic machinery that effectively allows the targeting of a lipid cargo to the lysosomes for degradation. A final confirmation of the process of "macrolipophagy" was obtained through studies in genetically modified mice that lacked autophagy within the liver (ATG7 conditional knockouts) (Singh et al. 2009a). These animals exhibit markedly enlarged livers that are lipid laden as demonstrated biochemically by increased TG and cholesterol levels (Singh et al. 2009a). In addition, Oil Red O staining of autophagy-deficient livers revealed marked increase in the number and size of lipid droplets when compared to control littermates (Singh et al. 2009a). Interestingly, the differences observed between controls and autophagy-ablated mice increased further when mice were fed a diet that provided 60% of calories in the form of fat. Although the lipophagic function of autophagy was initially demonstrated in cultured hepatocytes, in the liver, and in fibroblasts in culture (Singh et al. 2009a), it is now clear that macrolipophagy also regulates lipid stores in neurons (Martinez-Vicente et al. 2010).

Despite these exciting advances, a number of questions need to be addressed. It is still unclear how autophagosomes sequester lipid droplets given the enormous size of these lipid stores. An interesting possibility is that unconjugated LC3-I, which localizes on lipid droplets in basal condition, could acquire a phosphatidylethanolamine to form LC3-II that then generates a limiting membrane in situ sequestering the lipid droplet. Another integral question is to explore the possibilities of cross-talks between lysosomal lipases that degrade sequestered lipids and cytosolic neutral lipases. It is conceivable that smaller lipid droplets are engulfed completely by the autophagic apparatus, whereas larger lipid droplets are perhaps broken down into smaller droplets by autophagy that promotes further sequestration as well as increase the net lipid droplet surface area for efficient cytosolic neutral lipase activity.

### 2.2 *Intracellular Lipids and Regulation of Autophagy*

In contrast to the induction of autophagy observed following acute lipogenic stimulus, sustained lipogenic challenge or even acutely elevated abnormal lipid levels lead to compromised macrolipophagic function. Studies in cultured hepatocytes and mice fed a high-fat diet for prolonged periods have revealed that in these model systems there is a failure to mobilize intracellular lipid stores by macroautophagy (Singh et al. 2009a). This was reflected by electron microscopic evidence of decreased LC3-positive membranes on lipid droplets, as well as reduced areas of degeneration on lipid droplets following chronic lipogenic stimulus. Interestingly, a recent study by Koga et al. demonstrates that this autophagic defect following high-fat diet feeding is not limited to degradation of lipid cargo, but also alters degradation of all forms of autophagic cargo, including proteins, highlighting the fact that the primary defect lies in the autophagic apparatus (Koga et al. 2010). In this remarkable study, a dissection of the individual steps that mediate autophagy has demonstrated that high-fat diet feeding has no effect on induction or limiting membrane/autophagosome formation. In fact, the defect occurs at the level of autophagosome–lysosome fusion (Koga et al. 2010). The authors put forth an attractive possibility that alterations in membrane lipid composition following chronic high-fat diet feeding could potentially lead to decreased autophagosomal and lysosomal fusion. A crucial membrane lipid, cholesterol, plays an important role in membrane structure and function including membrane fusion events. Pharmacological agents that alter the concentration of autophagosomal and lysosomal membrane cholesterols resulted in altered membrane fusion events (Koga et al. 2010).

This dynamic cross-talk between autophagy and liver lipids has direct bearing to nonalcoholic fatty liver disease (NAFLD), the hepatic manifestation of the metabolic syndrome. The first step in the pathogenesis of NAFLD is the development of a fatty liver or hepatic steatosis that then predisposes to a "second hit," which is development of oxidative stress and inflammation leading to steatohepatitis and liver injury (Day and James 1998). In the human, it is conceivable that prolonged consumption of processed diets rich in fat, such as the western diet, can impair autophagy through the effects of lipids on autophagosome–lysosome fusion perpetuating a vicious circle of further hepatic fat accumulation. In addition, it remains to be explored whether compromised autophagy in the setting of fatty liver predisposes to the second hit that then leads to the development of steatohepatitis. In other words, is autophagy a central process that not only confers protection against steatosis but also blocks the development of steatohepatitis and end-stage liver disease in the setting of steatosis?

## 3 Autophagy and Adipose Tissue

The lipophagic function of autophagy demonstrated in the liver raised yet another critical question. Does autophagy play a similar role in the major fat-storing tissue in mammals, the adipose tissue?

## 3.1 *Autophagy and White Adipose Tissue*

As discussed in the previous section of this chapter, macrolipophagy has been demonstrated to be one of the mechanisms that mediate mobilization of intracellular lipid stores in hepatocytes, fibroblasts, and neurons. However, the function of autophagy in the dedicated fat-storing cell, the adipocyte, appears to be paradoxical to what has been observed in the aforementioned cell types. Recent studies by two independent groups have demonstrated a novel role of autophagy in the regulation of adipocyte differentiation and fat storage (Singh et al. 2009b; Zhang et al. 2009). In a study by Singh et al., inhibition of autophagy by RNA interference against ATG proteins in 3T3-L1 preadipocytes blocked the differentiation of preadipocytes into mature adipocytes (Singh et al. 2009b). This inhibition of differentiation of preadipocytes was associated with reduced expression of key adipogenic factors, CEBP-$\alpha$ and PPAR-$\gamma$. In addition, differentiation failure was associated with reduced TG storage, and at the molecular level by reduced levels of terminal differentiation markers, such as fatty acid binding protein-4 (FABP-4/aP-2), stearoyl-CoA desaturase, fatty acid synthase, and glucose transporter-4 in the autophagy-deficient adipocytes (Singh et al. 2009b).

White adipose tissue (WAT)-specific inhibition of autophagy *in vivo* by ablation of ATG7 not only reduced adipose tissue mass and differentiation but also imparted a remarkable brown adipose tissue (BAT)-like phenotype (Singh et al. 2009b). Reflective of this WAT to BAT transdifferentiation in the autophagy-deficient adipose tissue was histological evidence of smaller adipocytes with rounded nuclei that contained smaller, numerous multiloculated lipid droplets as opposed to control adipocytes with flattened nuclei containing a single large lipid droplet. In addition, morphometric analyses demonstrated increased mitochondrial content within the autophagy-deficient WAT (Singh et al. 2009b). To conclusively claim that inhibition of autophagy results in transdifferentiation of WAT into BAT-like tissue, it is imperative to demonstrate the expression of BAT-specific markers within the WAT. Indeed, immunoblotting demonstrated increased molecular markers of BAT; the PPAR-$\gamma$ transcriptional coactivator (PGC)-1$\alpha$ is crucial for brown adipogenesis and mitochondrial biogenesis, as well as presence of the specific BAT marker uncoupling protein-1 (UCP-1) (Singh et al. 2009b). One might argue that increased mitochondrial content could perhaps be a result of reduced mitophagy and not from increased biogenesis? However, if such a scenario were to be true, one would observe accumulation of predominantly dysfunctional mitochondria that would contribute to reduced oxidative function and cellular toxicity. Evidence against this possibility has been the demonstration of increased $\beta$-oxidation rates in adipose tissues of autophagy-ablated animals, regardless of whether mice were fed regular chow or high-fat diet. The physiological consequence of this dramatic alteration in WAT morphology including the acquisition of BAT-like properties is a remarkably lean mouse (Singh et al. 2009b) and maintained insulin sensitivity despite high-fat diet feeding (Singh et al. 2009b).

Despite these developments, it remains to be determined how absence of WAT-specific autophagy regulates adipose tissue differentiation and modulates the switch from WAT to a BAT-like phenotype. Recent studies have identified a number of proteins that regulate adipocyte biology (Lefterova and Lazar 2009). It could be possible that inhibition of autophagy blocks degradation of these critical adipocyte proteins that regulate adipogenesis and differentiation. Conversely, inhibition of autophagy could accumulate factors that impart BAT-like characteristics. An attractive candidate is PRD1-BF1-RIZ1 homologous domain containing 16 (PRDM-16), a transcriptional regulator that has recently been shown to promote a BAT-like phenotype through its interaction with PPAR-$\gamma$ and transcriptional upregulation of BAT-specific genes (Seale et al. 2008). A second possibility is a putative function of autophagy on adipocyte proliferation through effects on a subpopulation of adipose stromal vascular cells that are lineage − CD29+, CD34+, Sca-1+, and CD24+ and have been identified recently as WAT progenitor cells (Rodeheffer et al. 2008). However, since no differences have been observed in the relative percentage of these progenitor cells between control and autophagy-deficient fat pads (Singh et al. 2009b), a possibility exists that qualitative differences in progenitors may contribute to reduced adipose mass in WAT-specific autophagy null animals. It may also be possible that inhibition of autophagy perhaps alters the adipocyte metabolic milieu that exerts an indirect effect on fat storage. Despite all these possibilities, one cardinal question remains: how can one explain the differential roles of autophagy in the liver and adipose tissue? A plausible explanation for the divergent roles of autophagy in the liver and adipose tissue is the presence of an efficient lipolytic mechanism in the adipose tissue, i.e., the presence of the adipose TG lipase and hormone-sensitive lipase that act in tandem to mobilize adipose fat stores and liberate free fatty acids. Thus, a lipolytic function of autophagy in the adipose tissue akin to what has been described in the liver would prove to be redundant. Moreover, it can be envisioned that the mutually exclusive roles of autophagy in the liver and the adipose tissue could appear to be complementing each other to protect against fat storage in organs not functionally suited to store fat, such as liver and heart. In other words, the function of autophagy to promote adipose mass and differentiation allows efficient sequestration of excessive fat away from circulation, thus sparing the metabolic insult of fat accumulation in organs such as liver and heart.

## 3.2 *Autophagy and Brown Adipose Tissue*

The fact that inhibition of autophagy within WAT promotes the dramatic remodeling of WAT into a BAT-like phenotype does make one wonder: what would be the outcome of regulating autophagy within the interscapular BAT per se? The study by Singh et al. used a transgenic Cre mouse, the expression of which was driven by the aP2 promoter. The expression of aP2-driven Cre recombinase occurs in both WAT and BAT allowing the deletion of floxed ATG7 and hence autophagy in both adipose tissue compartments (Singh et al. 2009b). Interestingly, the inhibition of autophagy induced morphological and functional changes in BAT as well. This was suggested by the findings of a modest yet significant increase in the interscapular BAT mass in the autophagy null animals (Singh et al. 2009b). Histological sections of interscapular BAT from regular chow- and high-fat diet-fed autophagy null mice were remarkable for a reduction in the number of lipid droplets. Autophagy-deficient BATs demonstrated modest increases in UCP-1 and PGC-1α, as well as increased levels of mitochondrial markers cytochrome oxidase and cytochrome *c*, regardless of whether the mice were fed regular chow or a high-fat diet (Singh et al. 2009b). This was associated with a twofold increase in oxidative capacity of the autophagy-ablated BAT (Singh et al. 2009b). Thus, changes in BAT in animals lacking autophagy were consistent with those found in WAT but less significant because they occurred in the background of a fat depot already composed of brown adipocytes. A striking question is: why such a similar change, as observed in autophagy-null WAT, is occurring in autophagy-deficient BAT that is functionally and developmentally discrete from WAT? Is the fundamental mechanism increased retention of PRDM16 in both WAT and BAT that drives BAT gene expression? Only future studies will delineate the complex roles of autophagy in the regulation of adipocyte differentiation and fat storage and metabolism.

# 4 Conclusion

These exciting developments present autophagy as an organ-specific therapeutic target, which can be potentially manipulated to maintain energy homeostasis. In the liver, it is beneficial to upregulate autophagy that increases disposal of hepatic fat. In contrast, inhibition of adipose-specific autophagy enhances tissue oxidative capacity by imparting BAT-like properties that in turn promotes a lean insulin-sensitive phenotype. In addition, the effect of modulation of autophagy on the quantity and functioning of existing brown fat may have important implications for the development of novel therapeutic options for conditions that stem from increased adipose expansion such as the metabolic syndrome.

**Acknowledgments** This work was supported by National Institutes of Health grants from the National Institute of Diabetes and Digestive and Kidney Diseases to Dr. Ana Maria Cuervo (AMC) and Dr. Mark Czaja and from the National Institute of Aging and a Glenn Award to AMC; RS is supported by a NIH NIDDK K01 (DK087776-01) grant. The author thanks Dr. Ana Maria Cuervo for thought-provoking discussions and Dr. Susmita Kaushik for the critical reading of this manuscript.

# References

Beau I, Esclatine A, Codogno P (2008) Lost to translation: when autophagy targets mature ribosomes. Trends Cell Biol 18:311–314

Cuervo AM (2008) Autophagy and aging: keeping that old broom working. Trends Genet 24:604–612

Cuervo AM, Dice JF (1996) A receptor for the selective uptake and degradation of proteins by lysosomes. Science 273:501–503
Day CP, James OF (1998) Steatohepatitis: a tale of two "hits"? Gastroenterology 114:842–845
Esclatine A, Chaumorcel M, Codogno P (2009) Macroautophagy signaling and regulation. Curr Top Microbiol Immunol 335:33–70
He C, Klionsky DJ (2009) Regulation mechanisms and signaling pathways of autophagy. Annu Rev Genet 43:67–93
Jung CH, Jun CB, Ro SH, Kim YM, Otto NM, Cao J, Kundu M, Kim DH (2009) ULK-Atg13-FIP200 complexes mediate mTOR signaling to the autophagy machinery. Mol Biol Cell 20:1992–2003
Kabeya Y, Mizushima N, Ueno T, Yamamoto A, Kirisako T, Noda T, Kominami E, Ohsumi Y, Yoshimori T (2000) LC3, a mammalian homologue of yeast Apg8p, is localized in autophagosome membranes after processing. EMBO J 19:5720–5728
Koga H, Kaushik S, Cuervo AM (2010) Altered lipid content inhibits autophagic vesicular fusion. FASEB J 24:3052–3065
Kon M, Cuervo AM (2010) Chaperone-mediated autophagy in health and disease. FEBS Lett 584:1399–1404
Lefterova MI, Lazar MA (2009) New developments in adipogenesis. Trends Endocrinol Metab 20:107–114
Manjithaya R, Nazarko TY, Farre JC, Subramani S (2010) Molecular mechanism and physiological role of pexophagy. FEBS Lett 584:1367–1373
Martinez-Vicente M, Talloczy Z, Wong E, Tang G, Koga H, Kaushik S, de Vries R, Arias E, Harris S, Sulzer D, Cuervo AM (2010) Cargo recognition failure is responsible for inefficient autophagy in Huntington's disease. Nat Neurosci 13:567–576
Mizushima N (2009) Physiological functions of autophagy. Curr Top Microbiol Immunol 335:71–84
Petiot A, Ogier-Denis E, Blommaart EF, Meijer AJ, Codogno P (2000) Distinct classes of phosphatidylinositol 3′-kinases are involved in signaling pathways that control macroautophagy in HT-29 cells. J Biol Chem 275:992–998
Rodeheffer MS, Birsoy K, Friedman JM (2008) Identification of white adipocyte progenitor cells *in vivo*. Cell 135:240–249
Seale P, Bjork B, Yang W, Kajimura S, Chin S, Kuang S, Scime A, Devarakonda S, Conroe HM, Erdjument-Bromage H, Tempst P, Rudnicki MA, Beier DR, Spiegelman BM (2008) PRDM16 controls a brown fat/skeletal muscle switch. Nature 454:961–967
Singh R, Kaushik S, Wang Y, Xiang Y, Novak I, Komatsu M, Tanaka K, Cuervo AM, Czaja MJ (2009a) Autophagy regulates lipid metabolism. Nature 458:1131–1135
Singh R, Xiang Y, Wang Y, Baikati K, Cuervo AM, Luu YK, Tang Y, Pessin JE, Schwartz GJ, Czaja MJ (2009b) Autophagy regulates adipose mass and differentiation in mice. J Clin Invest 119:3329–3339
Tasdemir E, Maiuri MC, Tajeddine N, Vitale I, Criollo A, Vicencio JM, Hickman JA, Geneste O, Kroemer G (2007) Cell cycle-dependent induction of autophagy, mitophagy and reticulophagy. Cell Cycle 6:2263–2267
Tassa A, Roux MP, Attaix D, Bechet DM (2003) Class III phosphoinositide 3-kinase–Beclin1 complex mediates the amino acid-dependent regulation of autophagy in C2C12 myotubes. Biochem J 376:577–586
Tolkovsky AM (2009) Mitophagy. Biochim Biophys Acta 1793:1508–1515
Wei Y, Pattingre S, Sinha S, Bassik M, Levine B (2008) JNK1-mediated phosphorylation of Bcl-2 regulates starvation-induced autophagy. Mol Cell 30:678–688
Zhang Y, Goldman S, Baerga R, Zhao Y, Komatsu M, Jin S (2009) Adipose-specific deletion of autophagy-related gene 7 (atg7) in mice reveals a role in adipogenesis. Proc Natl Acad Sci USA 106:19860–19865

# Gene Co-Expression Modules and Type 2 Diabetes

**Alan D. Attie and Mark P. Keller**

**Abstract** Although most people with type 2 diabetes are obese, most obese people never develop diabetes. They are able to compensate for the insulin resistance that usually accompanies obesity by producing more insulin. We have replicated this obesity/diabetes dichotomy in mice by studying mouse strains that differ in diabetes susceptibility when made obese with the *Leptin*$^{ob}$ mutation; the C57BL/6 (B6) mice are resistant whereas the BTBR mice are susceptible. We have used this model system to search for genes that are causal for diabetes susceptibility. To find genes whose sequence variation leads to this phenotype, we have carried out positional cloning projects. To find genes that are responsive to genetic variation with respect to gene expression and are on pathways leading to diabetes, we have studied gene expression relative to gene variation; i.e., expression quantitative trait loci. We have used these data to generate network models, which incorporate gene loci, mRNA abundance, and other phenotypes. One of these networks is involved in the regulation of β-cell proliferation.

## 1 Gene Expression Profiling in Diabetes Research

The advent of microarray technology unleashed a torrent of surveys of gene expression in a wide array of human diseases, including obesity and diabetes (Attie and Keller 2010). Many key insights emerged from these studies, each from focusing on a particular tissue type. Below, we list selected examples from four different tissues.

In adipose tissue, the expression of genes involved in lipogenesis and adipogenesis is decreased in obese animals (Nadler et al. 2000; Soukas et al. 2000);

A.D. Attie (✉)
Department of Biochemistry, University of Wisconsin-Madison, Madison, WI 53706, USA
e-mail: attie@biochem.wisc.edu

W. Meyerhof et al. (eds.), *Sensory and Metabolic Control of Energy Balance*,
Results and Problems in Cell Differentiation 52,
DOI 10.1007/978-3-642-14426-4_5,  

(Dubois et al. 2006; Ducluzeau et al. 2001; Sewter et al. 2002). Microarray experiments revealed a strong increase in the expression of genes associated with inflammation (Lumeng et al. 2007a, b; Schenk et al. 2008; Weisberg et al. 2003; Xu et al. 2003). This is mainly due to an increase in the number of macrophages in adipose tissue rather than a change in the expression of these genes in adipocytes. It is now widely-accepted that obesity evokes an inflammatory response in adipose tissue. In fact, it has been suggested that the insulin resistance that accompanies obesity is a consequence of this inflammatory response.

In muscle, diabetes is associated with a coordinated, but subtle decrease in the expression of genes encoding enzymes involved in mitochondrial respiration (Mootha et al. 2003; Patti et al. 2003). Studies in muscle of old versus young individuals indicate that older people have a reduction in mitochondria number and muscle respiratory function (Petersen et al. 2003). It has been suggested that incomplete β-oxidation in muscle mitochondrial could produce byproducts of fatty acids that act to blunt insulin signaling (Koves et al. 2008). Paradoxically, several mutations that impair mitochondrial function are associated with improved insulin signaling (Hui et al. 2008; Pospisilik et al. 2007; Wredenberg et al. 2006). Thus, the relationship between mitochondrial function and insulin signaling is still not clearly understood.

In the liver, there is evidence of selective insulin resistance associated with insulin resistant states. The ability of insulin to suppress glucose production is diminished, but insulin is still able to induce *de novo* lipogenesis, a pathway largely driven by insulin's induction of SREBP-1c, a transcriptional activator of the entire pathway of lipogenesis. Recent studies attribute the selective insulin resistance to a branch in the insulin signaling pathway downstream of Akt (Li et al. 2010). Inhibitors of mTor block the induction of SREBP-1c, indicating that the induction of lipogenesis involves activation of mTor.

In β-cells, genes that have become prominent for their dysregulation in diabetes including thioredoxin interacting protein (*Txnip*) and *Arnt*/*Hif1β* are dysregulated in diabetes (Gunton et al. 2005; Shalev et al. 2002). These genes have been knocked out in mice. Knockout of *Txnip*, a gene induced by glucose, potently rescues several different animal models from diabetes (Chen et al. 2008). Knockout of *Arnt*, a gene suppressed by diabetes, leads to impaired insulin secretion (Gunton et al. 2005).

## 2 Coordinated Changes in Gene Expression Provide Physiological Clues

Although testing for differential gene expression in microarray studies is extremely useful, it has some limitations and is by no means the only information that can be extracted from microarray experiments. A major limitation stems from the fact that the technology enables the interrogation of tens of thousands of mRNA transcripts.

This results in a large multiple testing penalty and reduces the power to detect subtle changes in expression, which may be physiologically important.

One approach to overcome this problem is to reduce the dimensionality of the data. In one study, Mootha and co-workers showed that this could be done by curating the transcripts into just 146 categories, defined by physiological functions (Mootha et al. 2003). Then, they examined whether samples obtained from muscle biopsies of diabetic subjects contained an enrichment of transcripts of a particular functional category. Although the samples did not show statistically significant differential expression, there was very strong evidence that, as a group, genes encoding mitochondrial proteins were downregulated, albeit modestly.

Another approach to inferring altered physiological function from microarray experiments is to study the correlation structure of the data. This involves computing two-way correlation coefficients and then modifying the numbers so that they provide a larger separation of highly correlated from less highly correlated transcripts. These groups of transcripts are termed as "modules" (Zhang and Horvath 2005).

## 3 Genetics of Diabetes Susceptibility

Many physiological processes and organ systems contribute to diabetes-related phenotypes. These include, but are not limited to fuel partitioning, control of gluconeogenesis and lipogenesis, cytokines and their signaling pathways, and the function and growth dynamics of pancreatic β-cells. Thus it is no surprise that genome-wide association studies have identified many genes whose variants make relatively small contributions to diabetes risk. But, the one striking outcome of these studies is that a large proportion of the genes that have emerged are involved in some aspect of islet function. Many are genes related to cell proliferation, implying that control of β-cell proliferation plays a role in diabetes susceptibility in humans.

A problem at the heart of type 2 diabetes and central to our laboratory is to understand why some, but not all individuals who are obese develop type 2 diabetes. Our model system consists of two mouse strains that, when made obese with the *Leptin*$^{ob}$ mutation, differ in diabetes susceptibility. Thus, C57BL/6 (B6) mice are relatively diabetes-resistant while BTBR mice are develop severe diabetes when they become obese with the *Leptin*$^{ob}$ mutation (Clee and Attie 2006).

We bred the B6 and BTBR mouse strains carrying the *Leptin*$^{ob}$ mutation for two generations to generate an F2 population. This enabled us to map diabetes-related traits to polymorphic DNA markers distributed throughout the genome (Stoehr et al. 2000). We obtained quantitative trait loci (QTLs) for glucose and insulin, as well as many other phenotypes derived from measuring plasma analytes. As described below, these QTLs serve as anchor points for us to develop network models of diabetes causation.

## 4 Age, Obesity, and Strain Effects on Gene Expression and Diabetes

We surveyed gene expression in six tissues (adipose, liver, gastrocnemius and soleus muscle, hypothalamus, and islets) as a function of three variables; age, obesity, and mouse strain (Keller et al. 2008). Thus, we surveyed the tissues in the B6 and BTBR strains, lean or obese, at 4 weeks of age (when none of the mice are diabetic) and at 10 weeks (when the obese BTBR mice are diabetic). When we analyze the data, we hypothesize that changes in gene expression unique to the BTBR strain that occur at 4 weeks of age are potentially causative for diabetes whereas those changes occurring at only 10 weeks, after the onset of diabetes, are most likely consequences rather than causes of the disease (Fig. 1). This would also include gene expression changes observed in B6 but not in BTBR mice, or the reverse; changes observed in BTBR but not in B6, in response to obesity, since obesity unmasks the diabetes susceptibility of BTBR mice.

We used a method developed by Horvath to compute gene co-expression modules (Zhang and Horvath 2005). The connection strength between each transcript as well as the partial correlation coefficients between modules was computed, both within and between tissues. Altogether, we obtained 105 modules and created a connection map for each mouse strain (Fig. 2).

## 5 A Module Enriched in Cell Cycle Transcripts Predicts Diabetes

Of special interest, we identified a module in each of the tissues that was enriched for transcripts involved in cell cycle control (Fig. 3). In islets and adipose tissue, the expression of the genes in this module was coordinately increased by obesity. However, in islets, the induction by obesity was restricted to the B6 strain,

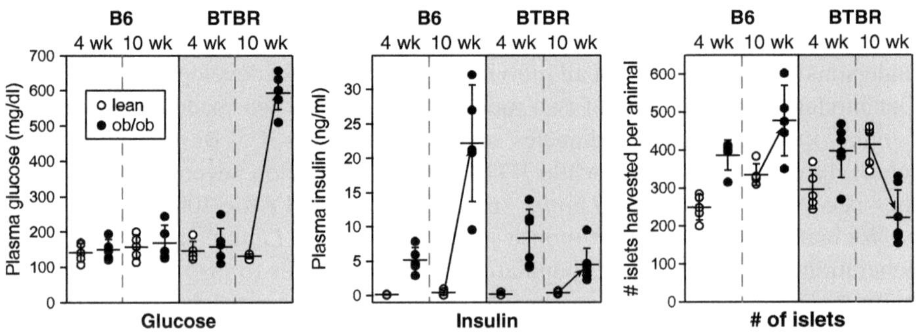


**Fig. 1** Effect of mouse strain, obesity, and age on serum glucose, insulin, and number of islets harvested. Clinical phenotypes are shown for 5–7 animals for each of the eight groups of mice used for study [Reprinted with permission from: Keller et al. (2008)]

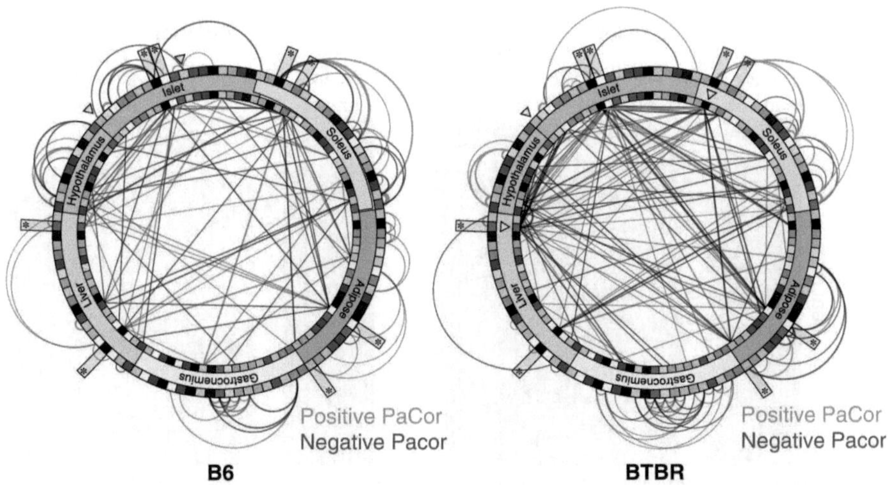


**Fig. 2** A gene–gene network model is distinct between B6 and BTBR mice. A gene–gene network was constructed based on the partial correlation (PaCor) between the strain-specific PC1 calculated between all modules identified in the six tissues profiled. Modules are illustrated as *bricks* along the inside and outside of the network wheels. Intertissue edges within the network are shown as lines connecting inside modules; intratissue edges are depicted as *arcs* connecting the outside modules. The cell cycle regulatory module in islet and those modules that form a direct connection to the cell cycle islet module are highlighted with *open arrow heads*. Network hot spots are indicated with *asterisks*. Line thickness is proportional to the magnitude of the PaCor, which ranged from 0.487 to 0.093 in B6 and from 0.303 to 0.086 in BTBR, for maximum and minimum respectively [Reprinted with permission from: Keller et al. (2008)]

suggesting that the diabetes susceptibility of the BTBR strain could be a consequence of a failure to respond to the insulin resistance brought about by obesity, by increasing the proliferation of pancreatic β-cells. In short, the gene expression patterns predicted that obesity stimulates β-cell expansion via β-cell replication in B6 mice (this was already known), but failed to do so in BTBR islets. In contrast to the islets, obesity induced the expression of the cell cycle module in adipose tissue in both B6 and BTBR mice.

To test the predictions of the cell cycle module, we carried out a direct measure of β-cell proliferation using a method developed by Hellerstein and co-workers (Macallan et al. 1998). The method involves administration of deuterium-labeled water to the mice for 2 weeks preceding sacrifice. The enrichment of deuterium in islet DNA is then determined by mass spectrometry and corrected to the enrichment in bone marrow DNA, to correct for differences in water consumption (especially important in diabetic mice) and to calculate the percentage of new islet cells (bone marrow cells undergo full turnover during the 2-week period). The direct measure of percentage of new cells showed ~2.5-fold increase in B6 islets and no increase in BTBR islets (Fig. 4), confirming the predictions of the gene expression studies. The deuterium-based proliferation measure in adipose tissue confirmed the prediction of the cell cycle modules; both strains showed increased adipocyte proliferation in response to obesity.

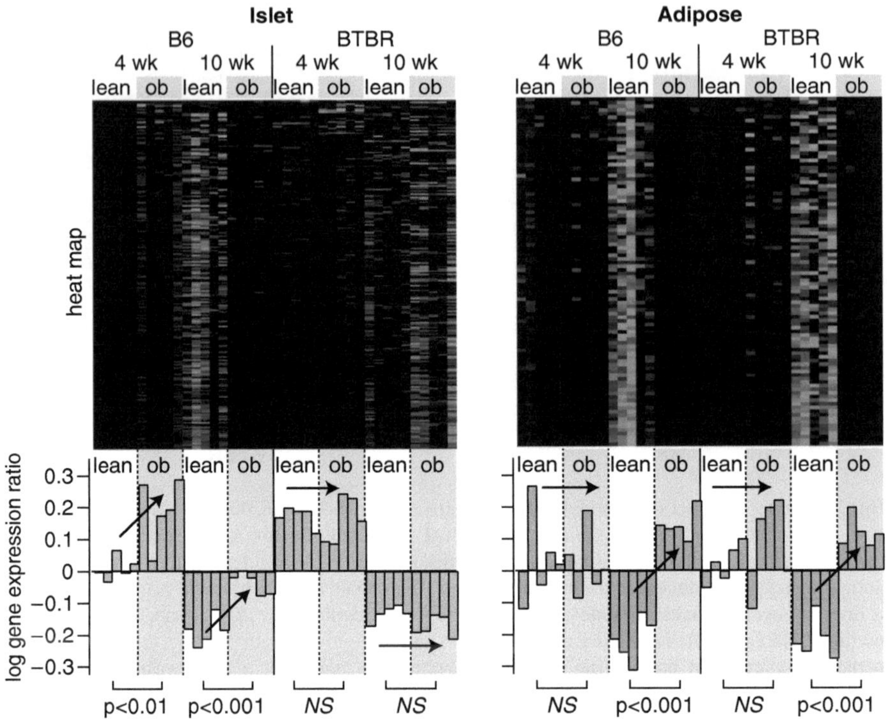


**Fig. 3** Co-expression modules enriched with cell cycle regulation accurately predict diabetes and obesity. Expression heat maps (**a**) and the first principal component (PC1) on $\log_{10}$ scale (**b**) of the cell cycle regulatory modules in islets (217 transcripts) and adipose (96 transcripts) are shown. *Bar plots* in **b** show the PC1 for individual mice and correspond to an expressed decrease for negative values and increased expression for positive values [Reprinted with permission from: Keller et al. (2008)]

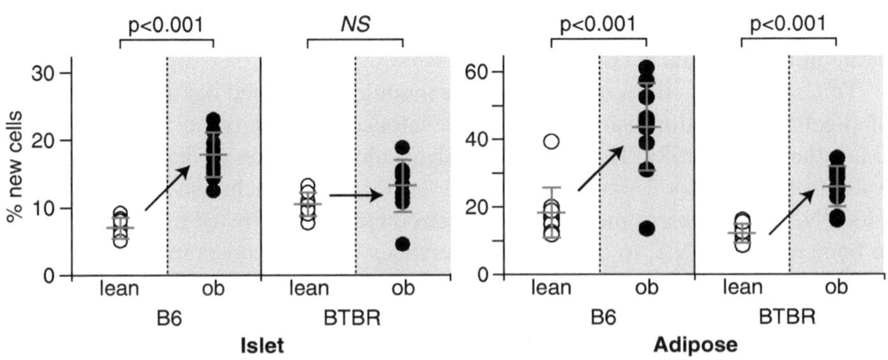


**Fig. 4** The percentage of new cells, derived from an in vivo measure of $^2$H incorporation into newly synthesized DNA, in islets and adipose tissue. Where significant obesity-dependent differences were observed, *p* values are shown. *Arrows* are used to show influence of obesity. *NS* not significant [Reprinted with permission from: Keller et al. (2008)]

## 6 Cell Cycle Module Shows Heritability

A major goal in diabetes research is to understand the pathways that regulate β-cell proliferation. It is possible that these pathways begin with some kind of communication between insulin's target tissues and the β-cells. This would ask how β-cells could "tune" their level of insulin output to the changes in insulin requirements, which are affected by changes in insulin sensing. Downstream of the signal to the β-cell are the pathways that stimulate the program of cell replication. Although the cell replication program itself is probably not unique in the β-cell, the pathways responsible for its regulation almost certainly are. We obtained some indication of this when we observed that obesity induces proliferation of adipocytes in both the B6 and BTBR strains but only induces proliferation of β-cells in the B6 strain.

To determine if there is heritability of the cell cycle gene expression module described above, we created a new F2 population ($n = 500$) of obese mice derived from the B6 and BTBR strains. An F2 population is one in which each animal is genetically unique, owing to the shuffling of chromosomal segments brought about by meiosis. Thus, if there are genomic regions that control groups of genes for a particular function, then that function should be stratified across the F2 population.

Remarkably, the islet and adipose cell cycle modules persisted in the F2 mice, despite the fact that all mice were obese and 10 weeks of age when sacrificed, suggesting that genetic differences between the strains were the primary driving force behind the coordinated gene expression changes reflected in the modules. We sorted the F2 population based solely on the expression of the islet cell cycle module. The F2 mice stratified into physiologically distinct subgroups. The mean glucose, insulin, number of islets harvested, C-peptide, and serum triglyceride levels were all significantly different between mice having high versus low expression of the cell cycle transcripts. As seen in Table 1, all five of these phenotypes were stratified.

In an F2 population, if phenotypes stratify, then there is a strong likelihood that there are distinct loci whose genotype correlates with the phenotype; i.e., linkage. Indeed, when we searched for "hot spots," i.e., genomic regions where many cell cycle transcripts linked, we found them on chromosomes 2, 4, 7, 10, and 17, with the largest number of transcripts mapping to chromosome 17.

When mRNA transcripts map to a particular region, they can do so in *cis* (proximal) or in *trans* (distal). When a transcript maps near the genomic region

**Table 1** Clinical phenotypes co-stratify with cell cycle module in F2 population

| Phenotype | Highest expression | Lowest expression | *p* value |
|---|---|---|---|
| Glucose (mg/dl) | 432 | 528 | $10^{-6}$ |
| Insulin (ng/ml) | 19 | 7 | $10^{-6}$ |
| Islets (#) | 257 | 179 | $10^{-10}$ |
| C-peptide (nM) | 5 | 3 | $10^{-10}$ |
| Triglycerides (mg/dl) | 170 | 277 | $10^{-7}$ |

where the gene encoding that transcript is physically located, then we denote that linkage as *cis*. The simplest example of such linkage is a polymorphism in a promoter region or a 3′-UTR that affects the rate of transcription or the stability of the mRNA transcript. When a transcript maps to a site that is distinct from the genomic region containing the gene encoding the transcript, we denote that linkage as *trans*. A simple example would be a transcript under the control of a transcript factor; i.e., a protein that is produced and can conceivably act at any gene locus. This analysis becomes a hypothesis generator when multiple transcripts map in *trans* to a single locus. Such mapping is consistent with the presence of one or more genes in the locus that coordinately regulate the abundance of many mRNA transcripts (e.g., a master regulator).

## 7 Future Prospects: Heritability Data Can Be Used to Create Network Models

One of the limitations of analysis of the high-volume data derived from the various types of – omics technologies is that it is quite easy to identify large numbers of correlated phenotypes, but it is much more difficult to define which of those correlations truly define causality. One of the reasons for this is that when two traits, designated "A" and "B," are correlated, A–B, it is impossible to infer whether or not A is causal for B (A → B), B is causal for A (A ← B), or if they are independent of one another and responding to another agent, X (A ← X → B). The equivalencies between these three models can be broken if there is genetic information; i.e., a genotype effect of a locus on A and B (Schadt et al. 2005). This occurs because, unlike the relationship between two phenotypes, the relationship between genotype and phenotype can only be one-way; i.e., genotype can influence phenotype, but phenotype cannot influence genotype (putting aside for the moment, epigenetics).

Our current efforts are focused on using linkage information to create causal network models for β-cell replication. A larger goal is to integrate multiple data sources to define networks involving mRNA, proteins, and metabolites. We previously performed a mixed phenotype network analysis where we combined metabolic QTLs with mRNA QTLs. This analysis predicted that glutamine would regulate several specific genes, including *Pck1*. We tested the model in cultured hepatocytes and showed that indeed, glutamine could strongly induce the expression of this gene.

The rapid development of technology to assess, with high throughput, protein abundance, posttranslational modifications of proteins, modifications on DNA, and metabolites, has already created a bottleneck at the level of data integration and analysis. Very few people possess the bioinformatic skills to perform the high-level computation necessary to distill data into an output that can be visualized while at the same time possessing the biological background to make meaning from the data. We and others are confronting these challenges by forming strong collaborative relationships with statisticians and computer scientists and by modifying the way we educate the next generation of biologists.

## References

Attie AD, Keller MP (2010) Physiological insights gained from gene expression analysis in obesity and diabetes. Annu Rev Nutr

Chen J et al (2008) Thioredoxin-interacting protein deficiency induces Akt/Bcl-xL signaling and pancreatic beta cell mass and protects against diabetes. FASEB J 22:3581–3594

Clee SM, Attie AD (2006) The genetic landscape of type 2 diabetes in mice. Endocr Rev 28:48–83

Dubois SG, Heilbronn LK, Smith SR, Albu JB, Kelley DE, Ravussin E (2006) Decreased expression of adipogenic genes in obese subjects with type 2 diabetes. Obesity (Silver Spring) 14:1543–1552

Ducluzeau PH et al (2001) Regulation by insulin of gene expression in human skeletal muscle and adipose tissue. Evidence for specific defects in type 2 diabetes. Diabetes 50:1134–1142

Gunton JE et al (2005) Loss of ARNT/HIF1beta mediates altered gene expression and pancreatic-islet dysfunction in human type 2 diabetes. Cell 122:337–349

Hui ST et al (2008) Txnip balances metabolic and growth signaling via PTEN disulfide reduction. Proc Natl Acad Sci U S A 105:3921–3926

Keller MP et al (2008) A gene expression network model of type 2 diabetes links cell cycle regulation in islets with diabetes susceptibility. Genome Res 18:706–716

Koves TR et al (2008) Mitochondrial overload and incomplete fatty acid oxidation contribute to skeletal muscle insulin resistance. Cell Metab 7:45–56

Li S, Brown MS, Goldstein JL (2010) Bifurcation of insulin signaling pathway in rat liver: mTORC1 required for stimulation of lipogenesis, but not inhibition of gluconeogenesis. Proc Natl Acad Sci U S A 107:3441–3446

Lumeng CN, Bodzin JL, Saltiel AR (2007a) Obesity induces a phenotypic switch in adipose tissue macrophage polarization. J Clin Invest 117:175–184

Lumeng CN, Deyoung SM, Bodzin JL, Saltiel AR (2007b) Increased inflammatory properties of adipose tissue macrophages recruited during diet-induced obesity. Diabetes 56:16–23

Macallan DC, Fullerton CA, Neese RA, Haddock K, Park SS, Hellerstein MK (1998) Measurement of cell proliferation by labeling of DNA with stable isotope-labeled glucose: studies in vitro, in animals, and in humans. Proc Natl Acad Sci U S A 95:708–713

Mootha VK et al (2003) PGC-1alpha-responsive genes involved in oxidative phosphorylation are coordinately downregulated in human diabetes. Nat Genet 34:267–273

Nadler ST, Stoehr JP, Schueler KL, Tanimoto G, Yandell BS, Attie AD (2000) The expression of adipogenic genes is decreased in obesity and diabetes mellitus. Proc Natl Acad Sci U S A 97:11371–11376

Patti ME et al (2003) Coordinated reduction of genes of oxidative metabolism in humans with insulin resistance and diabetes: potential role of PGC1 and NRF1. Proc Natl Acad Sci U S A 100:8466–8471

Petersen KF et al (2003) Mitochondrial dysfunction in the elderly: possible role in insulin resistance. Science 300:1140–1142

Pospisilik JA et al (2007) Targeted deletion of AIF decreases mitochondrial oxidative phosphorylation and protects from obesity and diabetes. Cell 131:476–491

Schadt EE et al (2005) An integrative genomics approach to infer causal associations between gene expression and disease. Nat Genet 37:710–717

Schenk S, Saberi M, Olefsky JM (2008) Insulin sensitivity: modulation by nutrients and inflammation. J Clin Invest 118:2992–3002

Sewter C et al (2002) Human obesity and type 2 diabetes are associated with alterations in SREBP1 isoform expression that are reproduced ex vivo by tumor necrosis factor-alpha. Diabetes 51:1035–1041

Shalev A et al (2002) Oligonucleotide microarray analysis of intact human pancreatic islets: identification of glucose-responsive genes and a highly regulated TGFbeta signaling pathway. Endocrinology 143:3695–3698

Soukas A, Cohen P, Socci ND, Friedman JM (2000) Leptin-specific patterns of gene expression in white adipose tissue. Genes Dev 14:963–980
Stoehr JP et al (2000) Genetic obesity unmasks nonlinear interactions between murine type 2 diabetes susceptibility loci. Diabetes 49:1946–1954
Weisberg SPD, McCann M, Desai M, Rosenbaum R, Leibel RL, Ferrante AW Jr (2003) Obesity is associated with macrophage accumulation in adipose tissue. J Clin Invest 116:1796–1808
Wredenberg A et al (2006) Respiratory chain dysfunction in skeletal muscle does not cause insulin resistance. Biochem Biophys Res Commun 350:202–207
Xu H et al (2003) Chronic inflammation in fat plays a crucial role in the development of obesity-related insulin resistance. J Clin Invest 112:1821–1830
Zhang B, Horvath S (2005) A general framework for weighted gene co-expression network analysis. Stat Appl Genet Mol Biol 4:Article17

# Role of Zinc Finger Transcription Factor *Zfp69* in Body Fat Storage and Diabetes Susceptibility of Mice

**Stephan Scherneck, Heike Vogel, Matthias Nestler, Reinhart Kluge, Annette Schürmann, and Hans-Georg Joost**

**Abstract** Type 2 diabetes is a polygenic disease resulting from a combination of different disease alleles reflecting obesity, insulin resistance, and hyperglycemia. Using a positional cloning strategy with different inbred strains of mice, we mapped a disease locus for obesity-associated diabetes on chromosome 4. We analyzed all genes in this region and identified distinct differences in the expression levels of the transcription factor *Zfp69*. The expression of this gene mediated diabetes progression in a leptin-deficient congenic mouse line. The animals developed a disease pattern of hyperglycemia, reduced gonadal fat mass, and increased plasma and liver triglycerides, resembling a potential defect in triglyceride storage. In order to elucidate the impact of the human ortholog of *Zfp69* in the development of type 2 diabetes, we tested its mRNA expression in human white adipose tissue. Consistent with the mouse data, mRNA-expression was significantly higher in diabetic subjects than in unaffected controls.

## 1 Introduction: Type 2 Diabetes as a Complex Genetic Disease

Type 2 diabetes is a complex disorder with a high degree of heritability (Das and Elbein 2006). The disease is the result of an interaction of adipogenic and diabetogenic alleles. Obese subjects do not necessarily develop diabetes; in contrast, the majority of type 2 diabetics are obese. This scenario indicates the need of the presence of both types of disease alleles, designated as "diabesity"-genes as well (NIH 1980; Shafrir 1992; Leiter and Chapman 1994; Joost 2008). Inbred strains of mice reflect the human situation powerfully. Dependent on the genetic background,

S. Scherneck (✉), H. Vogel, M. Nestler, R. Kluge, A. Schürmann, and H.-G. Joost
German Institute of Human Nutrition Potsdam-Rehbrücke, Arthur-Scheunert-Allee 114-116, 14558 Nuthetal, Germany
e-mail: scherneck@dife.de

W. Meyerhof et al. (eds.), *Sensory and Metabolic Control of Energy Balance*,
Results and Problems in Cell Differentiation 52,
DOI 10.1007/978-3-642-14426-4_6, 

obese animals are diabetes resistant or develop severe hyperglycemia and type 2 diabetes (Coleman 1978). The knowledge of the genetic constellation modifying the disease is an important option for the intervention of diabetes progression. Modern genotyping technologies have led to the identification of a huge number of single nucleotide polymorphisms (SNPs) associated with the disease. In the last 3 years genome-wide association studies (GWAs) identified tens of SNPs associated with type 2 diabetes and its related traits (Groves et al. 2006; Steinthorsdottir et al. 2007; Yasuda et al. 2008; Rung et al. 2009; Dupuis et al. 2010). However, in most cases, these SNPs were mapped in introns or outside genes. This fact makes it difficult to associate the mutation with changes in the function of the candidate gene (Joost 2009). A promising strategy for the identification of gene variants contributing to the disease is the genome-wide linkage analysis of outcross populations of inbred mouse strains. This method allows not only the identification of the disease gene but also the study of the functional mechanism underlying the complex trait.

## 2 Animal Models for the Study of Type 2 Diabetes

Several animal models exist for the study of complex diseases. For type 2 diabetes research rodents are of particular interest. Important examples in this field are both inbred and outbred strains of mice and rats. In addition, the sand rat (*Psammomys obesus*) and the spiny mouse (*Acomys cahirinus*) represent important models of diet-induced obesity and type 2 diabetes. These species show different metabolic responses to dietary effects of hyperglycemia (Shafrir et al. 2006). In contrast to these desert rodents, mice and rats offer a broad range of different characteristics. Mouse models for metabolic syndrome and type 2 diabetes research are available in a large number. This fact makes it difficult to select the “correct” model for the application of interest. Further, in most cases more than one strain is required for a sufficient modeling of the disease (Leiter 2009). Considering that type 2 diabetes is a polygenic disease, mouse models harboring single-gene mutations display limitations. To modulate the interaction between obesity, insulin resistance, and hyperglycemia so called “polygenic mouse models” are the organism of choice.

## 3 Polygenic Mouse Models for the Dissection of Type 2 Diabetes

It is easily comprehensible that obesity is the common feature of all mouse models of type 2 diabetes. Nevertheless, the genetic background of these mice is the important factor highlighting differences in the disease progression. A well established “polygenic” model is the New Zealand Obese (NZO) mouse. This strain develops a polygenic disease pattern of morbid obesity and insulin resistance. Further characteristics are an elevated blood pressure, serum cholesterol, and serum triglyceride levels (Ortlepp et al. 2000). Early events in the pathogenesis of the obese phenotype are hyperphagia, lower body temperature, and reduced running wheel activity (Jurgens et al. 2006). The identification of several loci in the NZO genome underlines the polygenic structure of the model. More than ten quantitative trait loci (QTL) for obesity were mapped in different crosses (Kluge et al. 2000; Reifsnyder et al. 2000; Taylor et al. 2001; Vogel et al. 2009). Hyperglycemia and the susceptibility to develop diabetes are also modulated by several loci. Some of these loci overlap with obesity QTL, such as the QTL *Nob3* on distal chromosome 1 (Vogel et al. 2009). The fact that obese and diabetic mice carry diabetes QTL is easily comprehensible. Interestingly, in different mouse crosses several diabetogenic alleles were contributed by lean and "healthy" strains. In an intercross of Non-obese Non-diabetic (NON) and NZO mice the loci *Nidd1* (chromosome 4) and *Nidd2* (chromosome 18) were identified as NON derived (Leiter et al. 1998). Further, a NZO cross with the lean Swiss Jim Lambert (SJL) strain identified the major diabetes QTL *Nidd/SJL* on distal chromosome 4. The diabetogenic allele was contributed by the SJL genome (Plum et al. 2000). Since in this cross the body weight of the progeny was strongly controlled by the obesity QTL *Nob1* on chromosome 5; a wide distribution of body mass was observed (Kluge et al. 2000). Dependent on their genotype for *Nidd/SJL*, lean animals did not differ in their susceptibility to diabetes. Animals with SJL-alleles for *Nidd/SJL* and obesity alleles for *Nob1* developed the highest blood glucose levels (Plum et al. 2002). In spite of the fact that the NZO strain carries diabetes genes itself, the variant responsible for the effects of *Nidd/SJL,* it additionally aggravates the diabetic phenotype (Fig. 1). This result underlines the relevance of the interaction of adipogenic and diabetogenic alleles.

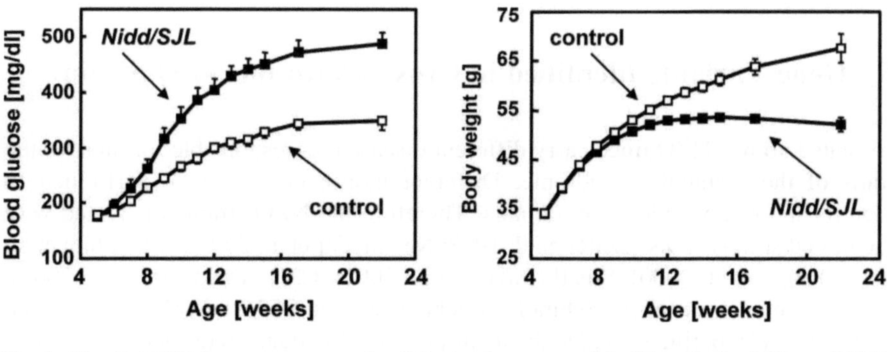


**Fig. 1** The *Nidd/SJL* locus aggravates the diabetic phenotype of the New Zealand Obese (NZO) mouse strain. The *Nidd/SJL* locus was transferred by breeding of recombinant-congenic mouse lines to the C57BL/6 strain. A twofold backcross to the NZO strain (N2) revealed hyperglycemia in both groups controls and *Nidd/SJL*-carrier. In contrast to the control group *Nidd/SJL*-carrier developed early hypoinsulinemia. In the course of the disease, the mice showed a significant reduction of weight gain based on the progressive type 2 diabetes

# 4 Positional Cloning of Diabesity Genes

The aim of the most linkage analyzes is the identification of the causal gene variant(s) of the disease. In so-called monogenic models, this goal is relative easy to achieve. For example, alleles causing obesity and diabetes of both the hormone leptin and its receptor *ob* and *db* were identified years ago (Bahary et al. 1990; Zhang et al. 1994; Chen et al. 1996; Chua et al. 1996; Lee et al. 1996). The discovery of these variants was possible without the knowledge of the full sequence of the mouse genome. In contrast, mapping of quantitative and complex traits is more complicated. Different loci in the genome interact with each other and with environmental components. So, in many cases it is not possible to separate a single locus from other loci. However, the first step in the dissection of complex traits is the mapping of the QTL responsible for the different characteristics of the disease (Korstanje and Paigen 2002). The linkage should satisfy the criteria of significance described by Lander and Kruglyak (1995) or by other groups (Shao et al. 2007). In a second step, linkage should be confirmed and the QTL region should be reduced to a critical region in which genes can be analyzed using molecular biological methods (Abiola et al. 2003). A promising strategy to achieve this goal is the generation of recombinant-congenic mouse lines (Demant and Hart 1986). This method enables the transfer of a chromosomal region from the affected strain to an unaffected control. If the trait controlled by the gene variant of interest is "stable," it is possible to define the critical QTL region via this method. Using interval-specific congenic strains it is possible to obtain critical segments of less than one centimorgan (Fehr et al. 2002). By the combination of sophisticated breeding strategies with suitable molecular biological methods several disease genes could be identified in the last years. Important examples are the successful positional cloning experiments of the obesity suppressor *Tbc1d1* (Chadt et al. 2008) and the type 2 diabetes genes *Sorcs1*, *Lisch-like*, and *Zfp69* (Clee et al. 2006; Dokmanovic-Chouinard et al. 2008; Scherneck et al. 2009).

# 5 Gene Variants Identified in Crosses with the NZO Strain

As stated above, NZO mice carry different disease loci responsible for most of the traits of the metabolic syndrome. Different mouse crosses were performed to identify the responsible gene variants. Therefore the NZO strain was mated with the inbred strains NON (Leiter et al. 1998; Reifsnyder et al. 2000), SJL (Plum et al. 2000; Kluge et al. 2000), Small (Taylor et al. 2001), C3H (Tsukahara et al. 2004), New Zealand Black (NZB; Schmolz unpublished), and C57 Black (B6; Vogel et al. 2009). These experiments led to the identification of variants in the genes *Lepr* (Igel et al. 1997), *Pctp* (Pan et al. 2006), *Abcg1* (Buchmann et al. 2007), *Nmur2* (Schmolz et al. 2007), *Tbc1d1* (Chadt et al. 2008), and *Zfp69* (Scherneck et al. 2009). Most of these variants show small effects which sum up in the polygenic phenotype of the model. In contrast, gene variants contributed by the lean strains (*Tbc1d1* and *Zfp69*) influence the disease pattern more dramatically.

## 6 Strategy to Narrow Down the *Nidd/SJL* Locus and Positional Cloning of *Zfp69*

*Zfp69* is located in the diabetes-susceptibility locus *Nidd/SJL* on distal chromosome 4. The locus contains approximately 600 genes. The strategy for the identification of the disease gene is based on the generation of recombinant-congenic mouse lines. In the first step, the susceptibility strain SJL was backcrossed to the diabetes-resistant C57BL/6 strain. This strategy led to generation of four different recombinant-congenic mouse lines differing in the size of the SJL interval introgressed in the C57BL/6 strain. As stated above, obesity is an essential factor for the development of type 2 diabetes. So, the lean recombinant-congenic lines were crossed with NZO mice to generate, an obese progeny. The recombinant congenic line representing the whole QTL interval of *Nidd/SJL* (genotype NZO/SJL) developed significantly higher blood glucose levels than the control line (genotype NZO/NZO). Two additional lines with smaller segments of the *Nidd/SJL* locus showed a similar phenotype. In contrast, the fourth mouse line showed similar blood glucose levels than the control line. Comparison of the chromosomal intervals represented by the recombinant-congenic mouse lines revealed a critical interval of the *Nidd/SJL* locus containing only ten genes (Fig. 2). The genes were sequenced and quantitative real-time polymerase chain reaction (qPCR) in liver, pancreas, skeletal muscle, and white adipose tissue was performed. The most striking result was obtained for the mRNA of the gene coding for Zinc finger protein 69 (*Zfp69*). In all investigated tissues, this gene was significantly higher expressed in the SJL strain than in C57BL/6 and NZO mice. In addition, a comparison of different tissues of recombinant-congenic mouse lines revealed that the detected differences in the expression of *Zfp69* is based on the genotype of the animals. Rapid amplification of cDNA ends (RACE) led to the identification of an aberrant exon in the strains C57BL/6 and NZO. This additional exon is based on the integration of an IAPLTR1a

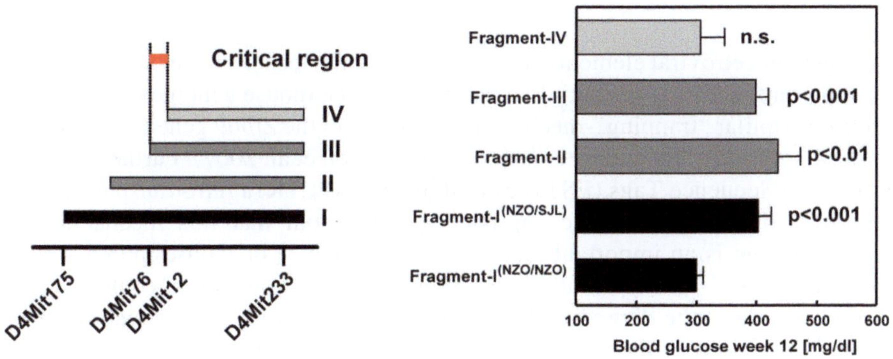


**Fig. 2** Identification of a small critical interval of the diabetes locus *Nidd/SJL*. Generation of different recombinant-congenic mouse lines and an additional reporter cross with the NZO strain defined a segment on chromosome 4 containing ten genes

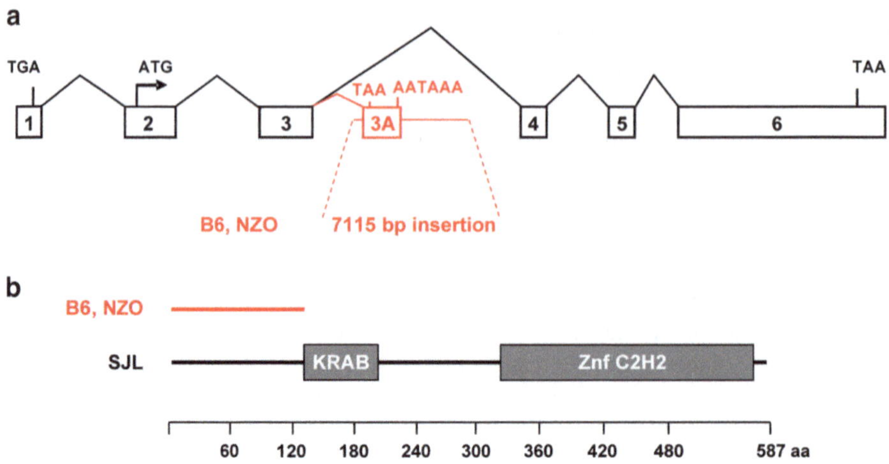


**Fig. 3** Disruption of the *Zfp69* gene by an IAPLTR1a retrotransposon in the strains B6 and NZO. (**a**) The transposon carried a splice-acceptor site and a polyadenylation signal, leading to a prematurely polyadenylated mRNA that lacked the coding sequence for (**b**) the Krüppel-associated box (KRAB) and Znf-C2H2 domains of *Zfp69*. SJL mice express the full-length mRNA

retrotransposon in intron 3 of the gene. This transposon carried a splice-acceptor site and a polyadenylation signal, leading to a prematurely polyadenylated mRNA that lacked the coding sequence for the Krüppel-associated box (KRAB) and Znf-C2H2 domains of *Zfp69* (Fig. 3). In conclusion, the aberrant mRNA of *Zfp69* in the mouse strains C57BL/6 and NZO seems to be "protective" in the development of hyperglycemia.

## 7 Genetic Variation Caused by the Insertion of Retroviral Elements

Endogenous retroviral elements such as IAP retrotransposons have been identified as important factors contributing to variations in the mouse genome (Zhang et al. 2008). A similar "trapping"-mechanism identified in the *Zfp69* gene was observed in the disruption of the mouse *Adamts13* gene (Zhou et al. 2007). Further aberrant Expressed Sequence Tags (ESTs) caused by the IAPLTR1a retrotransposon were identified by a bioinformatic approach, pointing out that this mechanism of gene silencing is an important factor in the divergence of mouse inbred strains (Scherneck et al. 2009). In addition to rodents, retrotransposons could play an important role in the diversity of the human genome. Sixty-five human disease-causing insertions of these "mobile elements" are known and this list is sure to grow (Goodier and Kazazian 2008). These findings should be considered in the design of future genotyping studies to investigate the human genome for disease-causing mutations.

## 8 The *Nidd/SJL–Zfp69* Phenotype Is Dependent on the Genetic Background

As described by Plum et al. (2002), the *Nidd/SJL* locus containing *Zfp69* led the NZO mice to hyperglycemia, hypoinsulinemia, and islet cell destruction based on their genetic background. In the course of the disease, the animals showed a significant reduction of weight gain based on the progressive type 2 diabetes. In conclusion, *Nidd/SJL–Zfp69* exacerbated the diabetic phenotype of the NZO strain (Fig. 1). Interestingly, the decompensatoric-hyperglycemic phenotype is in both the NZO parental strain and *Nidd/SJL–Zfp69* carriers on the NZO background, a gender-specific phenotype only present in male mice. The strong phenotype is a result of the combination of different diabetogenic alleles. The genetic background of NZO mice is diabetes sensitive, mainly controlled by the disease locus *Nob3* (Vogel et al. 2009). To study the effects of *Nidd/SJL–Zfp69* independent from the diabetes QTL of the NZO genome the whole QTL region of *Nidd/SJL–Zfp69* was transferred to the B6.V-$Lep^{ob}$ strain. The progeny of this breeding experiment was investigated dependent on the *Nidd/SJL–Zfp69* genotype. Animals carrying a *SJL* allele for the locus developed significantly higher blood glucose levels compared to animals with *B6* alleles (controls). The highest blood glucose levels of the *Nidd/SJL–Zfp69* carrier were observed between week 6 and week 8 of life (15–16 mmol/l). Up to an age of 28 weeks the animals compensated for hyperglycemia (blood glucose levels decreased to less than 9 mmol/l). This result underlines that the genetic background of the C57BL/6 strain, as present in B6.V-$Lep^{ob}$ mice, is diabetes resistant. In contrast to the genetic background of the NZO strain, the animals showed no islet cell destruction. Interestingly, the islets showed a significantly reduced immunoreactive insulin area than the islets of the control animals. Further, the carrier of the *Nidd/SJL–Zfp69* disease allele showed significantly less white adipose tissue and increased amounts of triglycerides in the liver. In addition, serum triglycerides were elevated in the *Nidd/SJL–Zfp69* disease allele carrier. These data suggest that the phenotype evoked by *Zfp69* is based on a defect in triglyceride storage leading to hepatosteatosis and hyperglycemia (Fig. 4). It is possible that the "islet" phenotype of the mice is independent of the changes in adipose tissue and needs further investigations.

## 9 *Zfp69* as a Member of KRAB-Containing Zinc-Finger Repressor Proteins (KRAB–ZNF)

*Zfp69* is expressed ubiquitously; e.g., in liver, skeletal muscle, white adipose tissue, pancreas, and testis. It is a member of a family of transcription factors that comprise of both the conserved KRAB and zinc finger DNA-binding domains (Urrutia 2003). Approximately, 300 members of this protein family are known but the functions of these transcription factors are poorly understood. Due to the fact that the *Nidd/SJL–Zfp69* disease allele carrier displayed smaller islets and significantly

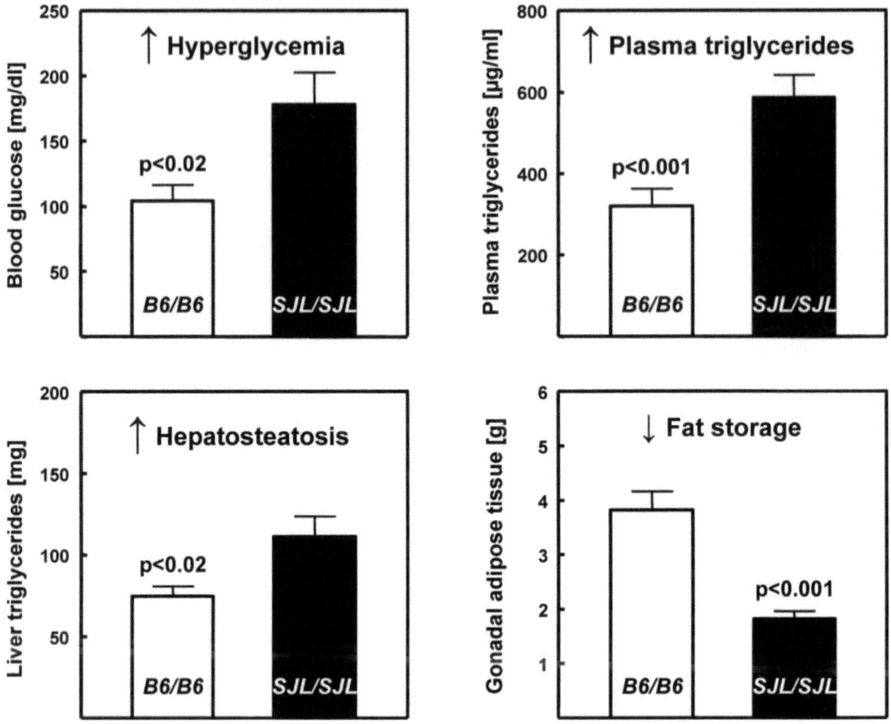


**Fig. 4** Reduced fat storage in carriers of *Nidd/SJL–Zfp69*. The transfer of the SJL variant of *Zfp69* to the B6.V-*Lep*$^{ob}$ strain led to a reduced storage of white adipose tissue, hepatosteatosis, hypertriglyceridemia, and hyperglycemia. On the *B6.V-Lep*$^{ob}$ background the animals showed no signs of β-cell failure

less white adipose tissue than control animals, *Zfp69* could be involved in proliferation processes in these tissues (Scherneck et al. 2009). From other members of this protein family, it is described that these transcription factors can suppress the MAPK signaling pathway (Liu et al. 2004; Cao et al. 2005; Huang et al. 2006) or the transcriptional activities of SRE and AP-1 (Liu et al. 2005). Further, the member of the KRAB–ZNF family, ZNF23, inhibits cell cycle progression (Huang et al. 2007). Additional experiments like ChIP on chip assays are required to identify the primary targets of *Zfp69* (Wang 2005).

## 10 The Role of the Human Ortholog of *Zfp69* in Type 2 Diabetes

*Zfp69* is conserved in human and mammals. The synthetic region of the mouse *Zfp69* locus maps to human chromosome 1p34. This locus contains two genes related to *Zfp69*, *ZNF642* and *ZNF643*. A comparison of the protein sequences revealed that *ZNF642* represents the human ortholog of *Zfp69*. To study the role of this gene in human type 2 diabetes both visceral adipose tissue and subcutaneous adipose tissue were investigated. The expression of the *ZNF642* mRNA was determined in diabetic and control individuals. In diabetic patients, *ZNF642* was significantly higher expressed than in healthy controls. In addition, there was a significant correlation of HbA1c levels with *ZNF642* mRNA. Subgroup analysis indicated that the correlation was significant in overweight but not in lean individuals.

## 11 Outlook

Within the last years, several candidate genes for type 2 diabetes were identified by positional cloning strategies in mice. The identification of *Zfp69* as a new target in the development of the disease underlines that these strategies are promising. At present, the target tissue responsible for the *Nidd/SJL–Zfp69* phenotype is not clearly identified. There is strong evidence that a reduced storage capacity of white adipose tissue is an early event in the pathogenesis of the disease. The presence of ectopic triglycerides stored in the liver suggests a higher degree of insulin resistance in this tissue. However, insulin resistance alone does not explain the morphometrical differences in the islet size of the animals expressing *Zfp69*. As a transcription factor, *Zfp69* could be involved in the regulation of β-cell proliferation in response to insulin resistance, but this hypothesis has to be proven.

## References

Abiola O, Angel JM, Avner P, Bachmanov AA, Belknap JK, Bennett B, Blankenhorn EP, Blizard DA, Bolivar V, Brockmann GA, Buck KJ, Bureau JF, Casley WL, Chesler EJ, Cheverud JM, Churchill GA, Cook M, Crabbe JC, Crusio WE, Darvasi A, de Haan G, Dermant P, Doerge RW, Elliot RW, Farber CR, Flaherty L, Flint J, Gershenfeld H, Gibson JP, Gu J, Gu W, Himmelbauer H, Hitzemann R, Hsu HC, Hunter K, Iraqi FF, Jansen RC, Johnson TE, Jones BC, Kempermann G, Lammert F, Lu L, Manly KF, Matthews DB, Medrano JF, Mehrabian M, Mittlemann G, Mock BA, Mogil JS, Montagutelli X, Morahan G, Mountz JD, Nagase H, Nowakowski RS, O'Hara BF, Osadchuk AV, Paigen B, Palmer AA, Peirce JL, Pomp D, Rosemann M, Rosen GD, Schalkwyk LC, Seltzer Z, Settle S, Shimomura K, Shou S, Sikela JM, Siracusa LD, Spearow JL, Teuscher C, Threadgill DW, Toth LA, Toye AA, Vadasz C, Van Zant G, Wakeland E, Williams RW, Zhang HG, Zou F (2003) The nature and identification of quantitative trait loci: a community's view. Nat Rev Genet 4:911–916

Bahary N, Leibel RL, Joseph L, Friedman JM (1990) Molecular mapping of the mouse db mutation. Proc Natl Acad Sci USA 87:8642–8646

Buchmann J, Meyer C, Neschen S, Augustin R, Schmolz K, Kluge R, Al-Hasani H, Jurgens H, Eulenberg K, Wehr R, Dohrmann C, Joost HG, Schurmann A (2007) Ablation of the cholesterol transporter adenosine triphosphate-binding cassette transporter G1 reduces adipose cell size and protects against diet-induced obesity. Endocrinology 148:1561–1573

Cao L, Wang Z, Zhu C, Zhao Y, Yuan W, Li J, Wang Y, Ying Z, Li Y, Yu W, Wu X, Liu M (2005) ZNF383, a novel KRAB-containing zinc finger protein, suppresses MAPK signaling pathway. Biochem Biophys Res Commun 333:1050–1059

Chadt A, Leicht K, Deshmukh A, Jiang LQ, Scherneck S, Bernhardt U, Dreja T, Vogel H, Schmolz K, Kluge R, Zierath JR, Hultschig C, Hoeben RC, Schurmann A, Joost HG, Al-Hasani H (2008) Tbc1d1 mutation in lean mouse strain confers leanness and protects from diet-induced obesity. Nat Genet 40:1354–1359

Chen H, Charlat O, Tartaglia LA, Woolf EA, Weng X, Ellis SJ, Lakey ND, Culpepper J, Moore KJ, Breitbart RE, Duyk GM, Tepper RI, Morgenstern JP (1996) Evidence that the diabetes gene encodes the leptin receptor: identification of a mutation in the leptin receptor gene in db/db mice. Cell 84:491–495

Chua SC Jr, Chung WK, Wu-Peng XS, Zhang Y, Liu SM, Tartaglia L, Leibel RL (1996) Phenotypes of mouse diabetes and rat fatty due to mutations in the OB (leptin) receptor. Science 271:994–996

Clee SM, Yandell BS, Schueler KM, Rabaglia ME, Richards OC, Raines SM, Kabara EA, Klass DM, Mui ET, Stapleton DS, Gray-Keller MP, Young MB, Stoehr JP, Lan H, Boronenkov I, Raess PW, Flowers MT, Attie AD (2006) Positional cloning of Sorcs1, a type 2 diabetes quantitative trait locus. Nat Genet 38:688–693

Coleman DL (1978) Obese and diabetes: two mutant genes causing diabetes-obesity syndromes in mice. Diabetologia 14:141–148

Das SK, Elbein SC (2006) The genetic basis of type 2 diabetes. Cellscience 2:100–131

Demant P, Hart AA (1986) Recombinant congenic strains – a new tool for analyzing genetic traits determined by more than one gene. Immunogenetics 24:416–422

Dokmanovic-Chouinard M, Chung WK, Chevre JC, Watson E, Yonan J, Wiegand B, Bromberg Y, Wakae N, Wright CV, Overton J, Ghosh S, Sathe GM, Ammala CE, Brown KK, Ito R, LeDuc C, Solomon K, Fischer SG, Leibel RL (2008) Positional cloning of "Lisch-Like", a candidate modifier of susceptibility to type 2 diabetes in mice. PLoS Genet 4:e1000137

Dupuis J, Langenberg C, Prokopenko I, Saxena R, Soranzo N, Jackson AU, Wheeler E, Glazer NL, Bouatia-Naji N, Gloyn AL, Lindgren CM, Magi R, Morris AP, Randall J, Johnson T, Elliott P, Rybin D, Thorleifsson G, Steinthorsdottir V, Henneman P, Grallert H, Dehghan A, Hottenga JJ, Franklin CS, Navarro P, Song K, Goel A, Perry JR, Egan JM, Lajunen T, Grarup N, Sparso T, Doney A, Voight BF, Stringham HM, Li M, Kanoni S, Shrader P, Cavalcanti-Proenca C, Kumari M, Qi L, Timpson NJ, Gieger C, Zabena C, Rocheleau G, Ingelsson E, An P, O'Connell J, Luan J, Elliott A, McCarroll SA, Payne F, Roccasecca RM, Pattou F, Sethupathy P, Ardlie K, Ariyurek Y, Balkau B, Barter P, Beilby JP, Ben-Shlomo Y, Benediktsson R, Bennett AJ, Bergmann S, Bochud M, Boerwinkle E, Bonnefond A, Bonnycastle LL, Borch-Johnsen K, Bottcher Y, Brunner E, Bumpstead SJ, Charpentier G, Chen YD, Chines P, Clarke R, Coin LJ, Cooper MN, Cornelis M, Crawford G, Crisponi L, Day IN, de Geus EJ, Delplanque J, Dina C, Erdos MR, Fedson AC, Fischer-Rosinsky A, Forouhi NG, Fox CS, Frants R, Franzosi MG, Galan P, Goodarzi MO, Graessler J, Groves CJ, Grundy S, Gwilliam R, Gyllensten U, Hadjadj S et al (2010) New genetic loci implicated in fasting glucose homeostasis and their impact on type 2 diabetes risk. Nat Genet 42:105–116

Fehr C, Shirley RL, Belknap JK, Crabbe JC, Buck KJ (2002) Congenic mapping of alcohol and pentobarbital withdrawal liability loci to a <1 centimorgan interval of murine chromosome 4: identification of Mpdz as a candidate gene. J Neurosci 22:3730–3738

Goodier JL, Kazazian HH Jr (2008) Retrotransposons revisited: the restraint and rehabilitation of parasites. Cell 135:23–35

Groves CJ, Zeggini E, Minton J, Frayling TM, Weedon MN, Rayner NW, Hitman GA, Walker M, Wiltshire S, Hattersley AT, McCarthy MI (2006) Association analysis of 6, 736 U.K. subjects provides replication and confirms TCF7L2 as a type 2 diabetes susceptibility gene with a substantial effect on individual risk. Diabetes 55:2640–2644

Huang X, Yuan W, Huang W, Bai Y, Deng Y, Zhu C, Liang P, Li Y, Du X, Liu M, Wang Y, Wu X (2006) ZNF569, a novel KRAB-containing zinc finger protein, suppresses MAPK signaling pathway. Biochem Biophys Res Commun 346:621–628

Huang C, Jia Y, Yang S, Chen B, Sun H, Shen F, Wang Y (2007) Characterization of ZNF23, a KRAB-containing protein that is downregulated in human cancers and inhibits cell cycle progression. Exp Cell Res 313:254–263

Igel M, Becker W, Herberg L, Joost HG (1997) Hyperleptinemia, leptin resistance, and polymorphic leptin receptor in the New Zealand obese mouse. Endocrinology 138:4234–4239

Joost HG (2008) Pathogenesis, risk assessment and prevention of type 2 diabetes mellitus. Obes Facts 1:128–137

Joost HG (2009) Genome-wide association studies, common variants and metabolic disease: do we now understand the genetic basis of human obesity and diabetes? Obes Metab-Milan 5:89–90, (ISSN: 1825–3865)

Jurgens HS, Schurmann A, Kluge R, Ortmann S, Klaus S, Joost HG, Tschop MH (2006) Hyperphagia, lower body temperature, and reduced running wheel activity precede development of morbid obesity in New Zealand obese mice. Physiol Genomics 25:234–241

Kluge R, Giesen K, Bahrenberg G, Plum L, Ortlepp JR, Joost HG (2000) Quantitative trait loci for obesity and insulin resistance (Nob1, Nob2) and their interaction with the leptin receptor allele (LeprA720T/T1044I) in New Zealand obese mice. Diabetologia 43:1565–1572

Korstanje R, Paigen B (2002) From QTL to gene: the harvest begins. Nat Genet 31:235–236

Lander E, Kruglyak L (1995) Genetic dissection of complex traits: guidelines for interpreting and reporting linkage results. Nat Genet 11:241–247

Lee GH, Proenca R, Montez JM, Carroll KM, Darvishzadeh JG, Lee JI, Friedman JM (1996) Abnormal splicing of the leptin receptor in diabetic mice. Nature 379:632–635

Leiter EH (2009) Selecting the "right" mouse model for metabolic syndrome and type 2 diabetes research. Methods Mol Biol 560:1–17

Leiter EH, Chapman HD (1994) Obesity-induced diabetes (diabesity) in C57BL/KsJ mice produces aberrant trans-regulation of sex steroid sulfotransferase genes. J Clin Invest 93:2007–2013

Leiter EH, Reifsnyder PC, Flurkey K, Partke HJ, Junger E, Herberg L (1998) NIDDM genes in mice: deleterious synergism by both parental genomes contributes to diabetogenic thresholds. Diabetes 47:1287–1295

Liu H, Zhu C, Luo J, Wang Y, Li D, Li Y, Zhou J, Yuan W, Ou Y, Liu M, Wu X (2004) ZNF411, a novel KRAB-containing zinc-finger protein, suppresses MAP kinase signaling pathway. Biochem Biophys Res Commun 320:45–53

Liu F, Zhu C, Xiao J, Wang Y, Tang W, Yuan W, Zhao Y, Li Y, Xiang Z, Wu X, Liu M (2005) A novel human KRAB-containing zinc-finger gene ZNF446 inhibits transcriptional activities of SRE and AP-1. Biochem Biophys Res Commun 333:5–13

NIH (1980) From the NIH: successful diet and exercise therapy is conducted in Vermont for "diabesity". JAMA 243:519–520

Ortlepp JR, Kluge R, Giesen K, Plum L, Radke P, Hanrath P, Joost HG (2000) A metabolic syndrome of hypertension, hyperinsulinaemia and hypercholesterolaemia in the New Zealand obese mouse. Eur J Clin Invest 30:195–202

Pan HJ, Agate DS, King BL, Wu MK, Roderick SL, Leiter EH, Cohen DE (2006) A polymorphism in New Zealand inbred mouse strains that inactivates phosphatidylcholine transfer protein. FEBS Lett 580:5953–5958

Plum L, Kluge R, Giesen K, Altmuller J, Ortlepp JR, Joost HG (2000) Type 2 diabetes-like hyperglycemia in a backcross model of NZO and SJL mice: characterization of a susceptibility locus on chromosome 4 and its relation with obesity. Diabetes 49:1590–1596

Plum L, Giesen K, Kluge R, Junger E, Linnartz K, Schurmann A, Becker W, Joost HG (2002) Characterisation of the mouse diabetes susceptibility locus Nidd/SJL: islet cell destruction, interaction with the obesity QTL Nob1, and effect of dietary fat. Diabetologia 45:823–830

Reifsnyder PC, Churchill G, Leiter EH (2000) Maternal environment and genotype interact to establish diabesity in mice. Genome Res 10:1568–1578

Rung J, Cauchi S, Albrechtsen A, Shen L, Rocheleau G, Cavalcanti-Proenca C, Bacot F, Balkau B, Belisle A, Borch-Johnsen K, Charpentier G, Dina C, Durand E, Elliott P, Hadjadj S, Jarvelin MR, Laitinen J, Lauritzen T, Marre M, Mazur A, Meyre D, Montpetit A, Pisinger C, Posner B, Poulsen P, Pouta A, Prentki M, Ribel-Madsen R, Ruokonen A, Sandbaek A, Serre D, Tichet J, Vaxillaire M, Wojtaszewski JF, Vaag A, Hansen T, Polychronakos C, Pedersen O,

Froguel P, Sladek R (2009) Genetic variant near IRS1 is associated with type 2 diabetes, insulin resistance and hyperinsulinemia. Nat Genet 41:1110–1115

Scherneck S, Nestler M, Vogel H, Bluher M, Block MD, Berriel Diaz M, Herzig S, Schulz N, Teichert M, Tischer S, Al-Hasani H, Kluge R, Schurmann A, Joost HG (2009) Positional cloning of zinc finger domain transcription factor Zfp69, a candidate gene for obesity-associated diabetes contributed by mouse locus Nidd/SJL. PLoS Genet 5:e1000541

Schmolz K, Pyrski M, Bufe B, Vogel H, Nogueiras R, Scherneck S, Nestler M, Zahn C, Ruschendorf F, Tschop M, Meyerhof W, Joost HG, Schurmann A (2007) Role of neuromedin-U in the central control of feeding behavior: a variant of the neuromedin-U receptor 2 contributes to hyperphagia in the New Zealand obese mouse. Obes Metab-Milan 3:28–37, (ISSN: 1825–3865)

Shafrir E (1992) Animal models of non-insulin-dependent diabetes. Diabetes Metab Rev 8:179–208

Shafrir E, Ziv E, Kalman R (2006) Nutritionally induced diabetes in desert rodents as models of type 2 diabetes: *Acomys cahirinus* (spiny mice) and *Psammomys obesus* (desert gerbil). ILAR J 47:212–224

Shao H, Reed DR, Tordoff MG (2007) Genetic loci affecting body weight and fatness in a C57BL/6J × PWK/PhJ mouse intercross. Mamm Genome 18:839–851

Steinthorsdottir V, Thorleifsson G, Reynisdottir I, Benediktsson R, Jonsdottir T, Walters GB, Styrkarsdottir U, Gretarsdottir S, Emilsson V, Ghosh S, Baker A, Snorradottir S, Bjarnason H, Ng MC, Hansen T, Bagger Y, Wilensky RL, Reilly MP, Adeyemo A, Chen Y, Zhou J, Gudnason V, Chen G, Huang H, Lashley K, Doumatey A, So WY, Ma RC, Andersen G, Borch-Johnsen K, Jorgensen T, van Vliet-Ostaptchouk JV, Hofker MH, Wijmenga C, Christiansen C, Rader DJ, Rotimi C, Gurney M, Chan JC, Pedersen O, Sigurdsson G, Gulcher JR, Thorsteinsdottir U, Kong A, Stefansson K (2007) A variant in CDKAL1 influences insulin response and risk of type 2 diabetes. Nat Genet 39:770–775

Taylor BA, Wnek C, Schroeder D, Phillips SJ (2001) Multiple obesity QTLs identified in an intercross between the NZO (New Zealand obese) and the SM (small) mouse strains. Mamm Genome 12:95–103

Tsukahara C, Sugiyama F, Paigen B, Kunita S, Yagami K (2004) Blood pressure in 15 inbred mouse strains and its lack of relation with obesity and insulin resistance in the progeny of an NZO/HILtJ × C3H/HeJ intercross. Mamm Genome 15:943–950

Urrutia R (2003) KRAB-containing zinc-finger repressor proteins. Genome Biol 4:231

Vogel H, Nestler M, Ruschendorf F, Block MD, Tischer S, Kluge R, Schurmann A, Joost HG, Scherneck S (2009) Characterization of Nob3, a major quantitative trait locus for obesity and hyperglycemia on mouse chromosome 1. Physiol Genomics 38:226–232

Wang JC (2005) Finding primary targets of transcriptional regulators. Cell Cycle 4:356–358

Yasuda K, Miyake K, Horikawa Y, Hara K, Osawa H, Furuta H, Hirota Y, Mori H, Jonsson A, Sato Y, Yamagata K, Hinokio Y, Wang HY, Tanahashi T, Nakamura N, Oka Y, Iwasaki N, Iwamoto Y, Yamada Y, Seino Y, Maegawa H, Kashiwagi A, Takeda J, Maeda E, Shin HD, Cho YM, Park KS, Lee HK, Ng MC, Ma RC, So WY, Chan JC, Lyssenko V, Tuomi T, Nilsson P, Groop L, Kamatani N, Sekine A, Nakamura Y, Yamamoto K, Yoshida T, Tokunaga K, Itakura M, Makino H, Nanjo K, Kadowaki T, Kasuga M (2008) Variants in KCNQ1 are associated with susceptibility to type 2 diabetes mellitus. Nat Genet 40:1092–1097

Zhang Y, Proenca R, Maffei M, Barone M, Leopold L, Friedman JM (1994) Positional cloning of the mouse obese gene and its human homologue. Nature 372:425–432

Zhang Y, Maksakova IA, Gagnier L, van de Lagemaat LN, Mager DL (2008) Genome-wide assessments reveal extremely high levels of polymorphism of two active families of mouse endogenous retroviral elements. PLoS Genet 4:e1000007

Zhou W, Bouhassira EE, Tsai HM (2007) An IAP retrotransposon in the mouse ADAMTS13 gene creates ADAMTS13 variant proteins that are less effective in cleaving von Willebrand factor multimers. Blood 110:886–893

# Metabolic Sensing in Brain Dopamine Systems

**Ivan E. de Araujo, Xueying Ren, and Jozélia G. Ferreira**

**Abstract** The gustatory system allows the brain to monitor the presence of chemicals in the oral cavity and initiate appropriate responses of acceptance or rejection. Among such chemicals are the nutrients that must be rapidly recognized and ingested for immediate oxidation or storage. In the periphery, the gustatory system consists of a highly efficient sensing mechanism, where distinct cell types express receptors that bind specifically to chemicals associated with one particular taste quality. These specialized receptors connect to the brain via dedicated pathways, the stimulation of which triggers stereotypic behavioral responses as well as neurotransmitter release in brain reward dopamine systems. However, evidence also exists in favor of the concept that the critical regulators of long-term nutrient choice are physiological processes taking place after ingestion and independently of gustation. We will appraise the hypothesis that organisms can develop preferences for nutrients independently of oral taste stimulation. Of particular interest are recent findings indicating that disrupting nutrient utilization interferes with activity in brain dopamine pathways. These findings establish the metabolic fate of nutrients as previously unanticipated reward signals that regulate the reinforcing value of foods. In particular, it suggests a role for brain dopamine reward systems as metabolic sensors, allowing for signals generated by the metabolic utilization of nutrients to regulate neurotransmitter release and food reinforcement.

---

I.E. de Araujo (✉), X. Ren, and J.G. Ferreira
The John B Pierce Laboratory and Department of Psychiatry, Yale University School of Medicine, 290 Congress Avenue, New Haven, CT 06519, USA
e-mail: IAraujo@jbpierce.org

W. Meyerhof et al. (eds.), *Sensory and Metabolic Control of Energy Balance*, Results and Problems in Cell Differentiation 52,
DOI 10.1007/978-3-642-14426-4_7, 

# 1 Introduction

The gustatory system allows the brain to monitor the presence of chemicals in the oral cavity and initiate acceptance or rejection responses accordingly. Among such chemicals are the potential fuels that must be rapidly recognized and ingested for later oxidation or storage. In fact, animals must continuously procure the substrates necessary to maintain cellular function from exogenous sources, that is, from food. A sensory system therefore evolved in which membrane receptors convey information on the presence of metabolic fuels in the oral cavity to brain circuits that control the initiation of ingestive behaviors.

However, the formation of long-term food preferences is a complex process, and different lines of evidence indicate that animals will fail to prefer foods conveying pleasant sensory cues if those are not ensued by postingestive, metabolic effects. How does the brain control food intake in such a way that previous associations between sensory properties and metabolic effects become a regulating factor during nutrient choice? One possibility is that a brain circuit exists where sensory and metabolic information will converge through independent pathways. In the following, after introducing the basic aspects related to the stimulation of brain reward systems by gustatory cues, we will review evidence in support of the idea that the midbrain dopamine system is one such candidate circuit. More specifically, we will evaluate current evidence that this brain neurotransmitter system is under the control of the nutritional state of the animal, with nutrient availability directly modulating neurotransmitter synthesis/release.

# 2 The Peripheral Gustatory System

The peripheral gustatory system corresponds to the anatomical substrate that links the sensory epithelium of the oral cavity to the first gustatory relay center in the brain. This includes a family of membrane proteins that function as chemical sensors, the epithelial cells hosting these sensors, and the neural afferents carrying information on sensor activation/cell depolarization to the brain. The oral chemosensory epithelia contain onion-shaped structures known as taste buds, which in turn typically host 50–100 taste receptor cells (Finger and Simon 2002). In mammals, TRCs are typically embedded in stratified epithelia and distributed throughout the oral cavity, more specifically expressed on tongue, palate, epiglottis, and esophagus (Finger and Simon 2002; Scott and Verhagen 2000; Spector and Travers 2005). The apical end of taste cells is exposed to the external environment of the oral cavity through a small opening in the epithelium called the taste pore, which is filled with microvilli. On the membranes of these microvilli are usually expressed different classes of receptors that function as oral chemosensors.

Proteins belonging to the G-protein-coupled receptor (GPCR) superfamily have been established as the receptors for sweet, L-amino acid and bitter tastants

(Adler et al. 2000; Chrandrashekar et al. 2000; Liu and Liman 2003; Max et al. 2001; Montmayeur et al. 2001; Mueller et al. 2005; Perez et al. 2003; Zhang et al. 2003; Zhao et al. 2003). On the other hand, the sensations associated with the other two primary tastants, namely sour and salty, are mediated by ion channels of the transient receptor potential (TRP) (Huang et al. 2006) and epithelial sodium channel (ENaC) (Kellenberger and Schild 2002) superfamilies. Sweet taste signaling is known to be mediated by heterodimeric GPCRs and specific downstream signaling elements. More precisely, the transduction of sweet tastants is mediated by the taste genes *Tas1r2* and *Tas1r3*, whose T1R2 and T1R3 products assemble to form the heterodimeric sweet receptor T1R2/T1R3 (Nelson et al. 2001; Zhang et al. 2003; Zhao et al. 2003). T1R2/T1R3 appears to be the one type of broadly tuned receptor that subserves detection of both natural sugars and artificial sweeteners, although it remains to be determined with exactitude whether these different classes of chemicals bind to different regions of the receptor (Nelson et al. 2001).

A similar mechanism mediates the recognition of L-amino acids via the *Tas1r1* and *Tas1r3* genes (Nelson et al. 2002). Accordingly, the transduction of most forms of L-amino acids (i.e., with the possible exception of aromatic L-amino acids) is primarily accomplished via the G-protein-coupled heterodimeric T1R1/T1R3 receptor (Nelson et al. 2002). T1R1/T1R3 receptors are broadly tuned to signal L-amino acids (Nelson et al. 2002; Zhao et al. 2003), although it has been proposed that the human form of the receptor is more narrowly tuned to glutamate or umami taste (Maruyama et al. 2006; Rong et al. 2005). Finally, the third class of tastants mediated by GPCRs includes bitter stimuli. Bitter taste is mediated by the *Tas2r* genes (Bufe et al. 2005), the products of which form homodimeric (i.e., containing two identical subunits) T2R receptors. Bitter T2R receptors have been found to be both necessary and sufficient for bitter taste transduction and perception (Mueller et al. 2005).

It appears that all signaling mechanisms downstream to taste GPCRs are shared by different classes of ligands. Upon receptor binding, taste GPCR signaling is supported by gustducin, a heterotrimeric taste G-protein whose α, β, and γ constituent units are α-gustducin (McLaughlin et al. 1992), Gβ3 and Gγ13 (Huang et al. 1999), respectively. A similar pattern seems to hold for signaling events occurring downstream to G-protein signaling. This includes the taste phospholipase PLCβ2 and the nonselective ionic taste channel TRPM5, the deletion of which induces severe impairments in – if not taste blindness for – sweet, umami, and bitter transduction (Zhang et al. 2003).

Finally, salty and sour taste sensations are mediated instead by ionic receptor channels. The ENaC, particularly its subunit ENaCα, mediates behavioral attraction to sodium chloride (Chandrashekar et al. 2010; Kretz et al. 1999). On the other hand, genetic and functional studies identified one member of the TRP superfamily, the polycystic kidney disease-like ion channel PKD2L1, as necessary for sour taste transduction (Huang et al. 2006; Ishimaru et al. 2006; LopezJimenez et al. 2006).

Upon receptor activation and taste cell depolarization, neural afferents originating from branches of cranial nerves innervate the basolateral aspect of taste cells and transmit to the brain information on the identity and quantity chemicals detected by the membrane taste receptors. The *chorda tympani* and the greater superior *petrosal* branches of the VIIth (facial) cranial nerve innervate TRCs present on the anterior tongue and palate, respectively (Danilova et al. 2002; Hanamori et al. 1988), such that information on chemosensory events occurring in the oral epithelium is transduced into electrical messages on its way to the brain.

## 3 The Central Gustatory System

Information derived from taste-responsive cranial nerves converges onto the rostral division of the nucleus tractus solitarius (rNTS) of the medulla (Hamilton and Norgren 1984), whereas the more caudal aspect of the NTS is targeted by visceral (vagal) afferent inputs that convey information on the physiological status of the gastrointestinal system (Travagli et al. 2006). From the rNTS, taste information ascends to further brain circuitries. In rodents, axonal fibers originating in this gustatory aspect of the nucleus of the solitary tract ascend ipsilaterally to the parabrachial nucleus (PBN), establishing this pontine structure surrounding the conjunctivum brachium as the second-order gustatory relay (Norgren and Leonard 1971, 1973; Norgren and Pfaffmann 1975). From PBN, a first ("dorsal") pathway projects to the parvicellular part of the ventroposterior medial nucleus of the thalamus (VPMpc), the taste thalamic nucleus (reviewed in Bermudez-Rattoni 2004). The second ("ventral") pathway includes direct projections from PBN to the central nucleus of the amygdala and lateral hypothalamus. Thalamic afferents then project to the primary gustatory cortex which is defined as the VPMpc cortical target.

## 4 Brain Dopamine Systems and Taste Reward

The role of brain dopamine systems in mediating food reward and encoding stimulus palatability has been well established (for a schematic representation of dopamine projections in the human brain, see Fig. 1). Dopamine antagonists attenuate the hedonic value of sweet-tasting nutrients, in that animals pretreated with either D1- or D2-type dopamine receptor antagonists behave toward high concentrations of sucrose solutions as if they were weaker than usual (Bailey et al. 1986; Geary and Smith 1985; Wise 2006; Xenakis and Sclafani 1981). Conversely, tasting palatable foods elevates dopamine levels in the nucleus accumbens (NAcc) of the ventral striatum (Hernandez and Hoebel 1988), a brain region largely implicated in food reinforcement (Kelley et al. 2005). In humans, striatal dopamine release directly correlates with the perceived hedonic value of food stimuli (Small et al. 2003). But is dopamine release induced by sweet palatability per se independently of carbohydrate metabolism? In fact, taste-elicited stimulation of the central dopamine systems seems to take place even in the absence of

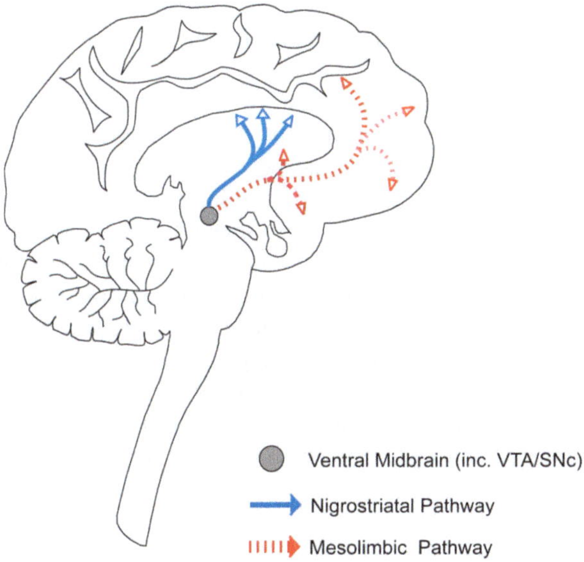


**Fig. 1** *Schematic representation of dopaminergic pathways in the human brain.* Analogous projections exist in both rodents and nonhuman primates. Dopaminergic neuronal cells residing in the midbrain locate within two regions of the ventral midbrain, the ventral tegmental area ("VTA") and the *pars compacta* of the *substantia nigra* ("SNc"). The mesolimbic pathway essentially consists of targets of VTA projections; these include mainly the *nucleus accumbens* of the ventral striatum, the amygdalar nuclei, the lateral hypothalamus and more medial, ventral parts of the frontal lobe. The mesolimbic pathway has been traditionally associated with the control of motivated behaviors and sensory reward processing, including food intake. In contrast, the nigrostriatal pathway consists of the targets reached by dopamine projections originating from SNc. This distinct, more dorsal, dopamine system has in turn been classically associated with motor control in general and initiation of voluntary movement in particular, being noticeable that SNc dopamine neurons seem to be preferentially targeted in Parkinson's disease. However, the functional distinctions between ventral and dorsal pathways have been blurred by recent findings showing a critical role for the dorsal striatum in feeding (see text for details)

intestinal nutrient absorption: In "sham-feeding" studies, where a catheter is implanted on the stomach wall to prevent nutrients from reaching the intestinal tract, accumbens' dopamine levels increase in proportion to the concentration of the sucrose solution used to stimulate the intraoral cavity (Hajnal et al. 2004).

It is therefore plausible to assume that the events leading to the stimulation of brain reward circuits via dopamine release initiate within the oral cavity, upon the activation of taste receptors. This implies that the dopamine release effect in accumbens associated with sweet taste stimulation must depend on the integrity of central taste relays conveying gustatory information to downstream brain circuits. Anatomically, whereas one group of projections from PBN reach the insular cortex via the taste thalamic relay (Norgren and Wolf 1975), a second, separate pathway reaches the amygdala, lateral hypothalamus, and the bed nucleus of the stria terminalis (Li et al. 2005; Norgren 1976). Thus, it has been shown that lesions to the PBN limbic, but not to the PBN thalamocortical, pathway blunt the dopaminergic response during intake of palatable tastants (Norgren and Hajnal 2005; Norgren et al. 2006).

More generally, the ability of pleasant sweet taste to directly stimulate brain dopamine systems appears consistent with the seemingly innate preferences to sweet taste in most species. In fact, it has been long shown that both deprived and nondeprived animals will not only avidly consume sweet solutions but also run through intricate mazes or incessantly press levers to obtain sweet rewards (Kare 1971). In humans, the innate attraction to sweetness is demonstrated by the reactions observed in children upon their first exposure to sugary solutions: newborns will immediately suck the solutions and produce characteristic facial expressions (Ganchrow et al. 1983). In rats, pups as young as 6 days old are strongly attracted to sweetness, given their robust intake responses to sweet compounds such as sucrose, lactose, and saccharin (Hall and Bryan 1981). It is also well established that other species show similar attractions to sweet compounds at early ages (Houpt et al. 1977).

## 5 Taste-Independent Attraction to Sugars

However, is sweetness perception required for animals to develop behavioral attractions to nutrients like sugars? This question can be more clearly addressed if the detection of certain sensory properties of distinct flavors is ablated during the experiments. One way to achieve this involves using genetically engineered animals lacking taste sensation. Following the lead open by flavor-nutrient conditioning paradigms (Sclafani and Vigorito 1987; Sclafani and Xenakis 1984), we have developed a conditioning protocol where mice are allowed to develop nutrient-specific preferences for sipper locations that had been previously associated with certain compounds (de Araujo et al. 2008). This study employed both wild-type and knockout mice lacking functional TRP channels M5 (Zhang et al. 2003). As was mentioned above, the TRPM5 ion channel is expressed in taste receptor cells (Perez et al. 2002) and is required for sweet, bitter, and amino acid taste signaling (Zhang et al. 2003). It was hypothesized in this study that sweet-blind *Trpm5* knockout mice would develop a preference for spouts associated with the presentation of sucrose solutions when allowed to detect the solutions' rewarding postingestive effects.

In fact, once the insensitivity of KO mice to the orosensory reward value of sucrose was established, we inquired whether a preference for sippers associated with caloric sucrose solutions could develop in water and food-deprived *Trpm5* knockout mice when they are allowed to form an association between a particular sipper in the test chamber and the postingestive effects produced by drinking from that sipper (de Araujo et al. 2008). This was accomplished in sweet taste-naïve animals by first determining the initial side preferences using a series of preliminary two-bottle tests where both sippers contained water and by exposing animals to conditioning sessions that consisted of daily 30 min free access to either water (assigned to the same side of initial bias) or sucrose (assigned to the opposite side) while access to the other sipper was blocked.

Confronting the behavioral data from both wild-type and knockout mice revealed no significant genotype × stimulus interactions, since during conditioning sessions both wild-type and knockout animals consumed significantly larger amounts of sucrose than water. In addition, during the postconditioning two-bottle tests, it was observed that both wild-type and knockout animals reversed their initial side-preference biases by drinking significantly more water from the sipper that during conditioning sessions had been associated with nutritive sucrose. Now, when the same experiments were run using the noncaloric sucrose-derived sweetener sucralose instead of sucrose, unlike the sucrose case, a significant genotype × stimulus interaction was detected since only wild-type animals consumed more sucralose than water during the conditioning sessions. Furthermore, during the two-bottle test sessions, conducted after conditioning to sucralose, knockout mice, likewise their wild-type counterparts, showed no preferences for sippers associated with the delivery of sucralose. Overall, these results provide evidence in favor of the hypothesis that postingestive effects can exert positive controls on ingestive (licking/swallowing) behaviors even in the absence of taste signaling or detection of distinct flavors. It is noticeable that even "taste-enabled" wild-type mice failed to develop preferences for sipper locations previously associated with sucralose delivery. This indicates that the mere presence of strong sweet taste input is not sufficient to induce long-term location preferences if unaccompanied by rewarding physiological effects.

## 6 On the Nature of the Postingestive Reward Signal

The above results raise the question of what would be the identity of the taste-independent reinforcement signal. Although there is little disagreement that postingestive factors produced by nutrients regulate food intake, much more controversial is the nature of the signal that acts on the brain as a postingestive reinforcer. Broadly speaking, the candidate signals could be classified into two major groups, related to pre- and postabsorptive postingestive events. The former group concerns those sensing mechanisms that occur before nutrient absorption but simultaneous to the arrival of nutrients to the gut. The latter group on the other hand refers to those events that occur following absorption, and nonexclusively includes a variety of signals such as fuel utilization metabolites and changes in plasma hormonal levels.

While it must be acknowledged that this is rather complex topic, currently available evidence points to a relatively weak role for preabsorptive signals in postingestive reinforcement. Regarding, for example, potential reinforcement signals generated in the stomach, early studies have shown that removing approximately 90% of the rat stomach results in gastrectomized rats being virtually as well motivated as controls during food-reinforced operant tasks (Tsang 1938). In addition, gastric vagotomy does not interfere with habitual feeding patterns in rats (Snowdon and Epstein 1970), and abdominal vagotomy did not interfere with flavor preferences conditioned by glucose-containing sugars (Sclafani and Lucas 1996).

Another possibility regards the presence of taste-like receptors in the gastrointestinal epithelium (Hofer et al. 1996). In fact, it has been shown that the taste signaling proteins α-Gustducin, T1R2, T1R3, as well as the taste ion channel TRPM5 are coexpressed in some mouse and human enteroendocrine cells (Bezençon et al. 2007; Margolskee et al. 2007). Accordingly, it is conceivable that gut cells could "taste" the contents of ingested foods and then convey (currently undetermined) signals to the brain that would function as behavioral reinforcers. Although this hypothesis deserves further attention, current evidence does not support a preponderant role for taste elements expressed in postingestive reinforcement. First we remind that *Trpm5* knockout mice do assimilate the differential physiological effects between sugar and sweetener solutions and develop sipper preferences accordingly (note that these mice do not express these taste channels anywhere in the body) (de Araujo et al. 2008). In addition, α-Gustducin knockout mice increase preferences for flavors associated with fat and sugar nutrients in ways that are comparable to wild-type mice (Sclafani et al. 2007), and T1R3 knockout mice do develop robust preferences for sucrose solutions (Zukerman et al. 2009). Overall, if nutrient sensors exist in the gut to mediate postingestive reinforcement signals, these sensors are unlikely to depend on taste receptor signaling.

Finally, one possibility remains regarding the potential role of gut-derived factors, such as the peptide hormones ghrelin or GLP-1, as postingestive reinforcers. Future experiments employing flavor-nutrient conditioning paradigms (Sclafani and Xenakis 1984) on the corresponding knockout models will contribute to resolve this issue. In any case, one potential argument against the involvement of such gut factors in postingestive reinforcement relates to their nonselectivity to rewarding compounds.

The second group of candidate postingestive reinforcers would consist of physiological signals generated postabsorption. We have recently assessed the potential role of metabolic signals in taste-independent nutrient selection by comparing the behavioral responses to glucose and L-amino acids in wild-type and *Trpm5* knockout mice (note that this channel mediates the tastes of sugars and L-amino acids in mice). Briefly, we have found that knockout mice, while displaying insensitivity to the tastes of glucose and L-serine during short-term tests, do develop a strong preference for glucose-associated compared to L-serine sippers over both conditioning and long-term sessions (Ren et al. 2010). In other words, animals will ingest higher quantities of, and develop preferences for, the carbohydrate glucose versus an isocaloric amino acid, even when unable to detect the distinctive flavor qualities intrinsic to each nutrient.

If the postingestive reinforcement signal is not exclusively explained by the actual number of calories ingested, which physiological cues could be playing such a role? Using indirect calorimetry measurements, we have found that these higher intake levels were closely associated with glucose oxidation levels, even more markedly than with increases in blood glucose. This finding points to a role for postabsorptive nutrient utilization in postingestive reinforcement that may be more important than any of the sensors detecting the presence of nutrients in either the gastrointestinal tract or bloodstream (Swithers and Davidson 2008). Now, these correlative measures do not necessarily provide direct evidence for the hypothesis that increased glucose intake is being primarily controlled by postabsorptive mechanisms. For example, nutrient-specific chemosensory signals and/or release of nutrient-controlling gut hormones (e.g., ghrelin, Tschop et al. 2000) might also have played a major role in shaping these behavioral responses. We have further clarified this issue by performing additional experiments where animals licked a water spout to obtain either glucose or serine infusions via a jugular catheter, thereby bypassing completely both the oral and gastrointestinal tracts. We have monitored the overall number of licks produced during the entire session and, as expected, we observed that mice licked significantly more times to water during glucose compared to during L-serine intravenous sessions. We conclude that the taste-independent differential responses to glucose and L-serine were not primarily accounted for by differential absorption rates or secretion of gut-derived factors, but rather by direct actions of nutrients on metabolism.

Altogether, the above findings provide support for the idea that postabsorptive mechanisms, independent of both oral and gastrointestinal sensing, ultimately mediate the higher intake and preference levels for glucose compared to amino acids. As we shall see below, a current working hypothesis relates to the possibility that nutrients providing sufficient fuel for brain metabolism may directly control neuronal activity in brain reward (dopamine) pathways, thereby reinforcing the previous behavioral sequences leading to its own intake. We will briefly describe evidence supporting a role for dopamine signaling in flavor-nutrient conditioning, and then provide evidence for a role for metabolic signals as regulators of brain dopamine pathways.

## 6.1 Brain Dopamine Signaling and Postingestive Conditioned Behaviors

Conditioned preferences for nutritive foods must ultimately depend on brain circuits that regulate ingestive behaviors. Among such circuits are those known to be involved in forming associations between unconditioned and conditioned reward stimuli. In fact, a role for dopamine signaling in flavor-nutrient conditioning was suggested by experiments where dopamine receptor antagonists were administered in the nucleus accumbens (the major dopamine target in ventral striatum strongly associated with feeding behaviors, see e.g., Baldo et al. (2005), Kelley et al. (2005). Rats treated with local infusions in nucleus accumbens with a D1-receptor antagonist displayed a dose-dependent reduction in intake of a flavor paired with intragastric infusions of glucose, compared to controls (Touzani et al. 2008). Interestingly, the effect of dopamine signaling antagonism on postconditioning preference tests was less compelling (Touzani et al. 2008). In any event, these results demonstrate that D1-like receptors in the nucleus accumbens are required for the acquisition, and possibly also for the expression, of glucose-conditioned flavor preferences.

On the other hand, our own findings suggest that the presence of taste or flavor stimulation is not required for the postingestive effects of foods to induce dopamine release in the nucleus accumbens. In fact, we observed by performing microdialysis in sweet-blind *Trpm5* knockout mice that sugar intake per se, independently of taste signaling, was sufficient to increase extracellular dopamine levels in the nucleus accumbens (de Araujo et al. 2008). More precisely, we first found in this study that the noncaloric sweetener sucralose produced significantly higher increases in dopamine levels in wild-type compared to knockout animals. These results are consistent with a role for dopamine signaling in accumbens derived from taste stimulation alone (Hajnal et al. 2004). However, when the same comparison was performed with respect to sucrose, no differences were found between the dopamine release levels in wild-type and knockout mice. In other words, while sweet taste stimulation without caloric content only produced significant increases in accumbal dopamine levels in wild-type, caloric sucrose evoked the same levels of dopamine increase in both wild-type and knockout mice. These results therefore strongly suggest that even in the absence of taste transduction and/or palatability, nutrient intake has the ability to induce measurable tonic increases in accumbens dopamine. Not only both palatability and postingestive factors seem to independently increase dopamine levels in brain reward circuits but also the nutrient-induced increases do not require the concomitant presence of flavor inputs, as had been suggested before (Di Chiara and Bassareo 2007). Therefore, the role played by dopamine signaling in postingestive reinforcement does not seem to be restricted to the formation of learned associations between orosensory and physiological signals; rather, nutrient availability may directly influence metabolic activity in cells present in this circuit, resulting in a simple, yet efficient metabolic-sensing reward machinery.

Earlier indications that postabsorptive signals might gain direct access to dopaminergic cells have been provided by the early work by Figlewicz and colleagues, who have shown that the functional forms of both insulin and leptin receptors (Figlewicz et al. 2003), as well as of some of their substrates (Pardini et al. 2006), are richly expressed in dopaminergic neurons of the substantia nigra compacta and ventral tegmental area regions of the midbrain. In addition, leptin receptors expressed in dopaminergic neurons of the midbrain were shown to be functional and to influence dopamine release (Fulton et al. 2006; Hommel et al. 2006). However, it is currently unknown whether brain leptin receptors play a role on postingestive reinforcement. In addition, the functional implications of insulin receptor expression in dopamine neurons have been little explored. Although it has been suggested that insulin infusions in the midbrain dopamine areas "decrease" the reward value of sucrose, since mice were found to reduce overall intake of sucrose solutions upon infusion (Figlewicz 2003; Figlewicz et al. 2006), this might simply imply that insulin receptor activation in midbrain dopamine areas provide the brain with a robust signal of caloric intake. However, a preponderant role for insulin as a postingestive reward signal is challenged by the findings that diabetic (hypoinsulinemic) rats do display normal responses in flavor-nutrient conditioning paradigms (Ackroff et al. 1997). Finally, Andrews et al. (2009) demonstrated that the orexigenic gut ghrelin promotes tyrosine hydroxylase gene expression in substantia nigra concomitantly to increasing dopamine concentration in striatum, raising the possibility that changes in ghrelin levels may modulate postingestive reinforcement.

## 6.2 *Glucose Metabolism and Dopamine Signaling*

Alternatively, a mechanism which would allow dopamine neurons to sense changes in physiological state refers to the possibility that these cells function as glucosensors, i.e., may change intracellular activity in response to the availability of extracellular glucose. More specifically, we directly addressed the possibility that dopamine neurons of the midbrain are sensitive to glucose utilization rates based on the finding (mentioned above) that higher intake levels of glucose compared to the nongluconeogenic amino acid L-serine were strongly associated with glucose oxidation levels (Ren et al. 2010).

The first step consisted of showing that intragastric infusions (i.e., completely bypassing the oral cavity) of glucose produce different effects on dopamine release compared to similar infusions of L-serine (see details in Fig. 2) (Ren et al. 2010). More precisely, intragastric infusions of glucose produced significantly higher levels of dopamine release in accumbens compared to isocaloric infusions of serine; we stress in particular the significant decreases in dopamine levels in accumbens following serine infusions, an effect that we relate to the lower levels of serine intake in KO animals. Furthermore, since dopamine signaling in dorsal striatum has also been implicated in feeding behavior (Sotak et al. 2005), we have in addition assessed the effects produced by intragastric infusions of glucose and serine on dorsal striatum dopamine levels. Whereas no significant decreases in dopamine levels were observed during L-serine infusions, significant increases were associated with glucose infusions. These microdialysis measures thus provided evidence that nutrient-specific dopamine release can be initiated upon direct stimulation of the gastrointestinal tract.

## 6.3 *Disrupting Glucose Metabolism Inhibits Dopamine Release in Dorsal Striatum*

To address the issue of whether the metabolism of glucose is relevant or not to nutrient-specific differences in dopamine release, we have designed another experiment where wild-type mice were fitted with a microdialysis probe in the striatum as

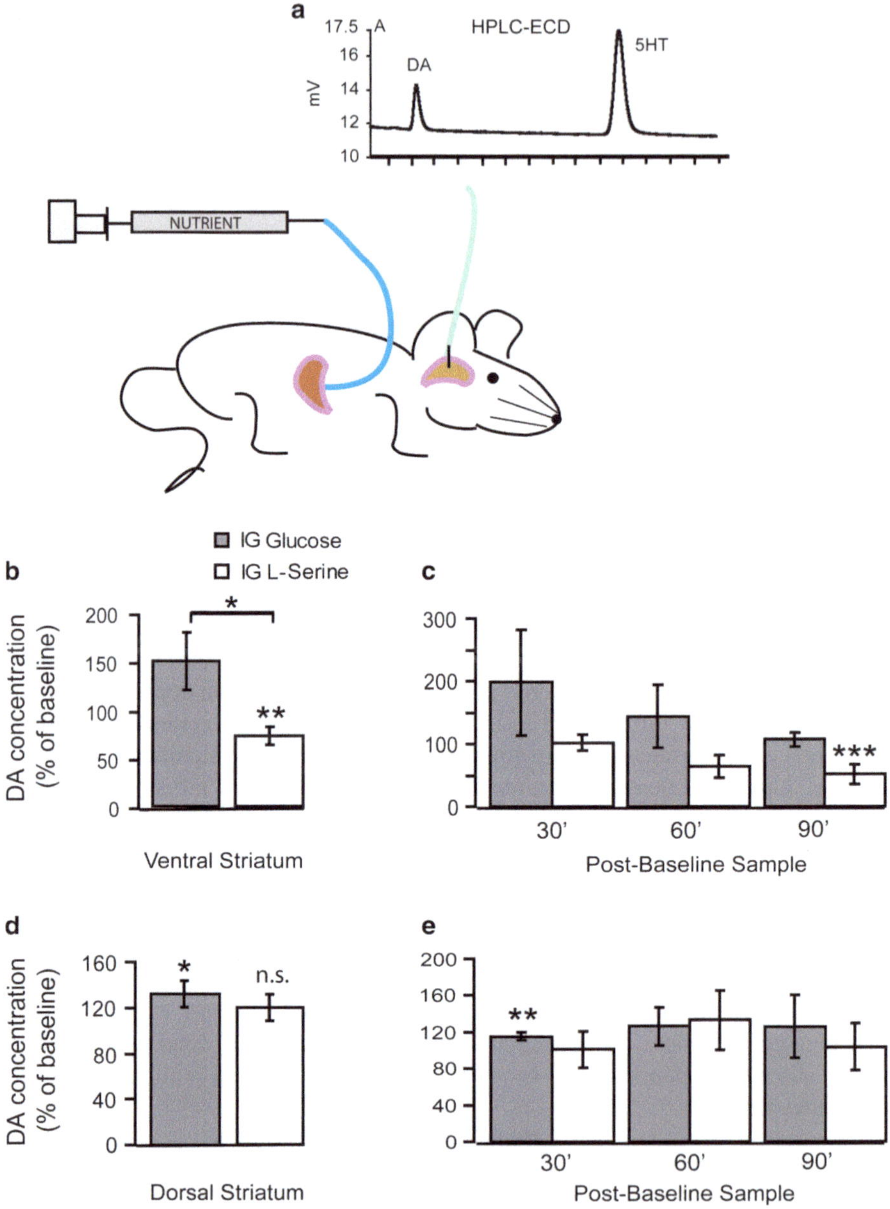


**Fig. 2** *Intragastric infusions of glucose and serine induce differential brain dopamine responses.* (**a**) Schematic showing the general arrangement where animals implanted with microdialysis probes are infused intragastrically with different nutrient levels, aiming at analyzing orosensory-independent monoamine release following food intake. (**b**) Overall percent changes in dopamine levels produced by glucose and serine intragastric infusions. In the nucleus accumbens, glucose infusions were associated with significantly higher levels of extracellular dopamine when directly compared to serine infusions (*two-sample *t* test $p < 0.03$). In addition, significant decreases in

well as with a jugular venous catheter. After measuring baseline dopamine levels, we have infused a bolus of the antimetabolic glucose analog, 2-deoxy-D-glucose (henceforth "2-DG") via the jugular catheter for 6 min and monitored dopamine levels for 1 h after the injection. This was then followed by an intravenous glucose infusion lasting for 10 min. We then reasoned that, if glucose metabolism is indeed relevant for the increased brain dopamine levels observed in striatum upon glucose ingestion, then an infusion of 2-DG should result in significant decreases in extracellular dopamine levels. In addition, such inhibitory effects of 2-DG on dopamine release must be reversed or at least attenuated by a subsequent intravenous glucose infusion that would contribute to restore normal rates of glucose oxidation.

In fact, intravenous infusions of 2-DG produced robust decreases in extracellular dopamine levels in striatum (34.7 ± 9.5% dopamine concentration of initial baseline). In addition, and also consistent with our earlier predictions, the subsequent intravenous infusions of glucose resulted in a reversal of this effect, with overall dopamine concentration levels reaching 74.5 ± 26.7% of the original baseline levels within 30 min. Importantly, we observed that glucose infusions produced a striking increase in striatal dopamine levels when the comparison is made with respect to the levels observed after 2-DG infusions. Therefore, glucose provision following inhibition of glucose utilization produces a strong relative increase in dopamine levels that are higher than those observed when no glucose utilization inhibition is employed.

## 6.4 Glucose Solutions Acquire Higher Reward Value When Contributing to Reinstate Glucose Oxidation

So far, we have been implicitly assuming that relative changes in extracellular dopamine levels reflect the reinforcing potency of a nutrient even in the absence of taste receptor signaling. Therefore, we are now forced to conclude that ingesting glucose following an injection of 2-DG must significantly increase the reward value of glucose compared to following a vehicle injection, even in sweet-insensitive *Trpm5* knockout animals. In other words, glucose solutions must be assigned a superior reward value when counteracting the effects of inhibiting glucose oxidation. We tested this hypothesis by measuring glucose intake in KO mice following

**Fig. 2** (continued) dopamine levels were observed following serine infusions (**one-sample $t$ test against 100% $p < 0.02$). (**c**) Whereas glucose infusions were consistently associated with increased dopamine levels across samples, relative decreases produced by serine infusions were marked in particular at the third sample (***$p < 0.04$). (**d**) In dorsal striatum, significant increases in dopamine levels were associated with glucose (*one-sample $t$ test against 100% $p = 0.009$), but not serine ($p < 0.09$), infusions. (**e**) Across samples, significant increases in dopamine levels were observed only during glucose infusions at 30 min after infusion onset (**one-sample $t$ test against 100% $p < 0.04$)

an intraperitoneal injection of either 2-DG or vehicle (saline). We have found in effect that, after 2-DG injections, knockout mice produced a significantly higher number of licks to glucose compared to after saline injections. Therefore, within 30 min of 2-DG administration, glucose acquired higher reward value when reinstating glucose oxidation levels, a finding that was entirely consistent with the dopamine measurements following 2-DG and glucose infusions described above.

## 7 Conclusion

Two conclusions may be drawn from the experimental evidence described in this chapter. First, ageusic mice unable to detect the orosensory properties of certain sugars or L-amino acids are nevertheless capable of developing nutrient-specific preferences based solely on physiological cues. Second, brain dopamine systems act as metabolic sensors, with a particular sensitivity to glucose oxidation rates. More generally, our results show that sugar-specific behavioral preferences and dopamine release will develop independently of sweetness or caloric value, while being regulated by glucose oxidation levels.

The above may contribute to explain the superior reinforcing value associated with glucose-containing sugars compared to all other classes of nutrients tested in flavor-nutrient conditioning paradigms (Ackoff 2009). In particular, these results may provide clues on whether the crucial reinforcing mechanism involves pre- or postabsorptive signals. In fact, it is conceivable that the stronger postingestive effects associated with glucose-containing sugars derive from the mere fact that neurons depend almost entirely on glucose and its derivatives for metabolic activity (Pellerin et al. 2007). This privileged access of glucose through the blood–brain barrier may underlie its superior reinforcing properties compared to other isocaloric nutrients. Future research must determine the mechanisms allowing the intracellular utilization of nutrients to regulate neurotransmitter release.

**Acknowledgments** We thank Prof Dietmar Richter for editorial assistance and Theddy Gonçalves for Fig. 1.

## References

Ackoff K (2009) Learned flavor preferences. The variable potency of post-oral nutrient reinforcers. Appetite 51:743–746

Ackroff K, Sclafani A, Axen KV (1997) Diabetic rats prefer glucose-paired flavors over fructose-paired flavors. Appetite 28:73–83

Adler E, Hoon MA, Mueller KL, Chrandrashekar J, Ryba NJP, Zucker CS (2000) A novel family of mammalian taste receptors. Cell 100:693–702

Andrews ZB, Erion D, Beiler R, Liu ZW, Abizaid A, Zigman J, Elsworth JD, Savitt JM, DiMarchi R, Tschoep M, Roth RH, Gao XB, Horvath TL (2009) Ghrelin promotes and

protects nigrostriatal dopamine function via a UCP2-dependent mitochondrial mechanism. J Neurosci 29:14057–14065
Bailey CS, Hsiao S, King JE (1986) Hedonic reactivity to sucrose in rats: modification by pimozide. Physiol Behav 38:447–452
Baldo BA, Alsene KM, Negron A, Kelley AE (2005) Hyperphagia induced by GABAA receptor-mediated inhibition of the nucleus accumbens shell: dependence on intact neural output from the central amygdaloid region. Behav Neurosci 119:1195–1206
Bermudez-Rattoni F (2004) Molecular mechanisms of taste-recognition memory. Nat Rev Neurosci 5:209–217
Bezençon C, le Coutre J, Damak S (2007) Taste-signaling proteins are coexpressed in solitary intestinal epithelial cells. Chem Senses 32:41–49
Bufe B, Breslin PAS, Kuhn C, Reed DR, Tharp CD, Slack JP, Kim UK, Drayna D, Meyerhof W (2005) The molecular basis of individual differences in phenylthiocarbamide and propylthiouracil bitterness perception. Curr Biol 15:322–327
Chandrashekar J, Kuhn C, Oka Y, Yarmolinsky DA, Hummler E, Ryba NJ, Zuker CS (2010) The cells and peripheral representation of sodium taste in mice. Nature 464:297–301
Chrandrashekar J, Mueller KL, Hoon MA, Adler E, Feng L, Guo W, Zucker CS, Ryba NJP (2000) T2Rs function as bitter taste receptors. Cell 100:703–711
Danilova V, Danilov Y, Roberts T, Hellekant G (2002) Sense of taste of the common marmoset: recordings from the chorda tympani and glossopharyngeal nerves. J Neurophys 88: 579–594
de Araujo IE, Oliveira-Maia AJ, Sotnikova TD, Gainetdinov RR, Caron MG, Nicolelis MA, Simon SA (2008) Food reward in the absence of taste receptor signaling. Neuron 57:930–941
Di Chiara G, Bassareo V (2007) Reward system and addiction: what dopamine does and doesn't do. Curr Opin Pharmacol 7:69–76
Figlewicz DP (2003) Insulin, food intake, and reward. Semin Clin Neuropsychiatry 8:82–93
Figlewicz DP, Evans SB, Murphy J, Hoen M, Baskin DG (2003) Expression of receptors for insulin and leptin in the ventral tegmental area/substantia nigra (VTA/SN) of the rat. Brain Res 964:107–115
Figlewicz DP, Bennett JL, Naleid AM, Davis C, Grimm JW (2006) Intraventricular insulin and leptin decrease sucrose self-administration in rats. Physiol Behav 89:611–616
Finger TE, Simon SA (2002) The cell biology of lingual epithelia. In: Finger TF, Silver WL, Restrepo D (eds) The Neurobiology of Taste and Smell, Wiley-Liss, New York, pp. 287–314
Fulton S, Pissios P, Manchon RP, Stiles L, Frank L, Pothos EN, Maratos-Flier E, Flier JS (2006) Leptin regulation of the mesoaccumbens dopamine pathway. Neuron 51:811–822
Ganchrow JR, Steiner JE, Daher M (1983) Neonatal facial expressions in response to different qualities and intensities of gustatory stimuli. Infant Behav Dev 6:473–484
Geary N, Smith GP (1985) Pimozide decreases the positive reinforcing effect of sham fed sucrose in the rat. Pharmacol Biochem Behav 22:787–790
Hajnal A, Smith GP, Norgren R (2004) Oral sucrose stimulation increases accumbens dopamine in the rat. Am J Physiol Regul Integr Comp Physiol 286:R31–R37
Hall WG, Bryan TE (1981) The ontogeny of feeding in rats: IV. Taste development as measured by intake and behavioral responses to oral infusions of sucrose and quinine. J Comp Physiol Psychol 95:240–251
Hamilton RB, Norgren R (1984) Central projections of gustatory nerves in the rat. J Comp Neurol 222:560–577
Hanamori T, Miller IJ Jr, Smith DV (1988) Gustatory responsiveness of fibers in the hamster glossopharyngeal nerve. J Neurophysiol 60:478–498
Hernandez L, Hoebel BG (1988) Food reward and cocaine increase extracellular dopamine in the nucleus accumbens as measured by microdialysis. Life Sci 42:1705–1712
Hofer D, Puschel B, Drenckhahn D (1996) Taste receptor-like cells in the rat gut identified by expression of alpha-gustducin. Proc Natl Acad Sci USA 93:6631–6634

Hommel JD, Trinko R, Sears RM, Georgescu D, Liu ZW, Gao XB, Thurmon JJ, Marinelli M, DiLeone RJ (2006) Leptin receptor signaling in midbrain dopamine neurons regulates feeding. Neuron 51:801–810
Houpt KA, Houpt TR, Pond WG (1977) Food intake controls in the suckling pig: glucoprivation and gastrointestinal factors. Am J Physiol 232:E510–514
Huang L, Shanker YG, Dubauskaite J, Zheng JZ, Yan W, Rosenzweig S, Spielman AI, Max M, Margolskee RF (1999) Ggamma13 colocalizes with gustducin in taste receptor cells and mediates IP3 responses to bitter denatonium. Nat Neurosci 2:1055–1062
Huang AL, Chen X, Hoon MA, Chandrashekar J, Guo W, Trankner D, Ryba NJ, Zuker CS (2006) The cells and logic for mammalian sour taste detection. Nature 442:934–938
Ishimaru Y, Inada H, Kubota M, Zhuang H, Tominaga M, Matsunami H (2006) Transient receptor potential family members PKD1L3 and PKD2L1 form a candidate sour taste receptor. Proc Natl Acad Sci U S A 103:12569–12574
Kare MR (1971) Comparative study of taste. In: Beidler LM (ed) Handbook of sensory physiology. Springer, Berlin, pp 278–292
Kellenberger S, Schild L (2002) Epithelial sodium channel/degenerin family of ion channels: a variety of functions for a shared structure. Physiol Rev 82:735–767
Kelley AE, Schiltz CA, Landry CF (2005) Neural systems recruited by drug- and food-related cues: studies of gene activation in corticolimbic regions. Physiol Behav 86:11–14
Kretz O, Barbry P, Bock R, Lindemann B (1999) Differential expression of RNA and protein of the three pore-forming subunits of the amiloride-sensitive epithelial sodium channel in taste buds of the rat. J Histochem Cytochem 47:51–64
Li CS, Cho YK, Smith DV (2005) Modulation of parabrachial taste neurons by electrical and chemical stimulation of the lateral hypothalamus and amygdala. J Neurophys 93:1183–1196
Liu D, Liman ER (2003) Intracellular Ca2+ and the phospholipid PIP2 regulate the taste transduction ion channel TRPM5. Proc Natl Acad Sci U S A 100:15160–15165
LopezJimenez ND, Cavenagh MM, Sainz E, Cruz-Ithier MA, Battey JF, Sullivan SL (2006) Two members of the TRPP family of ion channels, Pkd1l3 and Pkd2l1, are co-expressed in a subset of taste receptor cells. J Neurochem 98:68–77
Margolskee RF, Dyer J, Kokrashvili Z, Salmon KS, Ilegems E, Daly K, Maillet EL, Ninomiya Y, Mosinger B, Shirazi-Beechey SP (2007) T1R3 and gustducin in gut sense sugars to regulate expression of Na + -glucose cotransporter 1. Proc Natl Acad Sci U S A 104(38):15075–15080
Maruyama Y, Pereira E, Margolskee RF, Chaudhari N, Roper SD (2006) Umami responses in mouse taste cells indicate more than one receptor. J Neurosci 26:2227–2234
Max M, Shankar G, Huang L, Rong M, Liu Z, Campagne F, Weinstein H, Damak S, Margolskee RF (2001) Tas1r3, encoding a new candidate taste receptor, is allelic to the sweet responsiveness locus. Sac Nat Gen 28:58–63
McLaughlin SK, McKinnon PJ, Margolskee RF (1992) Gustducin is a taste-cell specific G protein closely related to transducins. Nature 357:563–569
Montmayeur JP, Liberlis SD, Matsunami H, Buck L (2001) A candidate taste receptor gene near a sweet taste locus. Nat Neurosci 4:492–498
Mueller KL, Hoon MA, Erlenbach I, Chandrashekar J, Zuker CS, Ryba NJP (2005) The receptors and coding logic for bitter taste. Nature 434:225–229
Nelson G, Hoon MA, Chandrashekar J, Ryba NJP, Zuker CS (2001) Mammalian sweet taste receptors. Cell 106:381–390
Nelson G, Chandrashekar J, Hoon MA, Feng L, Zhao G, Ryba NJP, Zucker CS (2002) An amino-acid taste receptor. Nature 726:1–4
Norgren R (1976) Taste pathways to hypothalamus and amygdala. J Comp Neurol 166:17–30
Norgren R, Hajnal A (2005) Taste pathways that mediate accumbens dopamine release by sapid sucrose. Physiol Behav 84:363–369
Norgren R, Leonard CM (1971) Taste pathways in rat brainstem. Science 173:1136–1139
Norgren R, Leonard CM (1973) Ascending central gustatory pathways. J Comp Neurol 150:217–238
Norgren R, Pfaffmann C (1975) The pontine taste area in the rat. Brain Res 91:99–117

Norgren R, Wolf G (1975) Projections of thalamic gustatory and lingual areas in the rat. Brain Res 92:123–129

Norgren R, Hajnal A, Mungarndee SS (2006) Gustatory reward and the nucleus accumbens. Physiol Behav 89:531–535

Pardini AW, Nguyen HT, Figlewicz DP, Baskin DG, Williams DL, Kim F, Schwartz MW (2006) Distribution of insulin receptor substrate-2 in brain areas involved in energy homeostasis. Brain Res 1112:169–178

Pellerin L, Bouzier-sore AK, Aubert S, Serres S, Merle M, Costalat R, Magistretti PJ (2007) Activity-dependent regulation of energy metabolism by astrocytes: an update. Glia 55: 1251–1262

Perez CA, Huang L, Rong M, Kozak JA, Preuss AK, Zhang H, Max M, Margolskee RF (2002) A transient receptor potential channel expressed in taste receptor cells. Nat Neurosci 5:1169–1176

Perez CA, Margolskee RF, Kinnamon SC, Ogura T (2003) Making sense with TRP channels: store-operated calcium entry and the ion channel Trpm5 in taste receptor cells. Cell Calcium 33:541–549

Ren X, Ferreira JG, Zhou L, Shammah-Lagnado SJ, Yeckel CW, De Araujo IE (2010) Nutrient selection in the absence of taste receptor signaling. J Neurosci 30:8012–8023

Rong M, He W, Yasumatsu K, Kokrashvili Z, Perez CA, Mosinger B, Ninomiya Y, Margolskee RF, Damak S (2005) Signal transduction of umami taste: insights from knockout mice. Chem Sens 30:i33–i34

Sclafani A, Lucas F (1996) Abdominal vagotomy does not block carbohydrate-conditioned flavor preferences in rats. Physiol Behav 60:447–453

Sclafani A, Vigorito M (1987) Effects of SOA and saccharin adulteration on polycose preference in rats. Neurosci Biobehav Rev 11:163–168

Sclafani A, Xenakis S (1984) Sucrose and polysaccharide induced obesity in the rat. Physiol Behav 32:169–174

Sclafani A, Zukerman S, Glendinning JI, Margolskee RF (2007) Fat and carbohydrate preferences in mice: the contribution of {alpha}-Gustducin and Trpm5 taste signaling proteins. Am J Physiol Regul Integr Comp Physiol 293:R1504–R1513

Scott TR, Verhagen JV (2000) Taste as a factor in the management of nutrition. Nutrition 16:874–885

Small DM, Jones-Gotman M, Dagher A (2003) Feeding-induced dopamine release in dorsal striatum correlates with meal pleasantness ratings in healthy human volunteers. Neuroimage 19:1709–1715

Snowdon CT, Epstein A (1970) Oral and intragastric feeding in vagotomized rats. J Comp Physiol Psychol 71:59–67

Sotak BN, Hnasko TS, Robinson S, Kremer EJ, Palmiter RD (2005) Dysregulation of dopamine signaling in the dorsal striatum inhibits feeding. Brain Res 1061:88–96

Spector AC, Travers SP (2005) The representation of taste quality in the mammalian nervous system. Behav Cogn Neurosci Rev 4:143–191

Swithers SE, Davidson TL (2008) A role for sweet taste: calorie predictive relations in energy regulation by rats. Behav Neurosci 122:161–173

Touzani K, Bodnar R, Sclafani A (2008) Activation of dopamine D1-like receptors in nucleus accumbens is critical for the acquisition, but not the expression, of nutrient-conditioned flavor preferences in rats. Eur J Neurosci 27:1525–1533

Travagli RA, Hermann GE, Browning KN, Rogers RC (2006) Brainstem circuits regulating gastric function. Annu Rev Physiol 68:279–305

Tsang YC (1938) Hunger motivation in gastrectomized rats. J Comp Physiol Psychol 26:1–17

Tschop M, Smiley DL, Heiman ML (2000) Ghrelin induces adiposity in rodents. Nature 407:908–913

Wise RA (2006) Role of brain dopamine in food reward and reinforcement. Philos Trans R Soc Lond B Biol Sci 361:1149–1158

Xenakis S, Sclafani A (1981) The effects of pimozide on the consumption of a palatable saccharin–glucose solution in the rat. Pharmacol Biochem Behav 15:435–442

Zhang Y, Hoon MA, Chandrashekar J, Mueller KL, Cook BWD, Zucker CS, Ryba NJ (2003) Coding of sweet, bitter, and umami tastes: different receptor cells sharing similar signaling pathways. Cell 112:293–301

Zhao GQ, Zhang Y, Hoon MA, Chandrashekar J, Erienbach I, Ryba NJP, Zuker CS (2003) The receptors for mammalian sweet and umami taste. Cell 115:255–266

Zukerman S, Touzani K, Margolskee R, Sclafani A (2009) Role of olfaction in the conditioned sucrose preference of sweet-ageusic T1R3 knockout mice. Chem Sens 35:685–694

# Oral and Extraoral Bitter Taste Receptors

**Maik Behrens and Wolfgang Meyerhof**

**Abstract** The role of bitter taste receptors has changed considerably over the past years. While initially considered to have predominantly, or even exclusively, gustatory functions, numerous recent reports addressed nongustatory actions of TAS2Rs. One site of extraoral bitter taste receptor expression is the respiratory system. It was demonstrated that bitter taste receptors are located in the nasal respiratory epithelium, as well as in ciliated cells of lung epithelium, where they affect respiratory functions in response to noxious stimuli. Another site of TAS2R gene expression is the gastrointestinal tract. Here, bitter compounds are suspected to regulate via activation of TAS2Rs metabolic and digestive functions.

The present article focuses on general pharmacological features and signal transduction components of mammalian TAS2Rs and summarizes current knowledge on Tas2r gene function in respiratory and gastrointestinal systems on the expense of a detailed description of gustatory bitter taste perception, which has been the subject of recent reviews.

## 1 Introduction

The sense of taste fulfills the important function to evaluate the quality and nutritional value of food prior to its ingestion. Each of the five basic taste qualities – sweet, sour, umami, salty, and bitter – reports about a specific attribute of the consumed food present in the oral cavity. Carbohydrates and amino acids elicit sweet and umami taste sensations, respectively, indicating a high caloric content, salty signals the presence of sodium ions important for our body's electrolyte

M. Behrens (✉) and W. Meyerhof
Department Molecular Genetics, German Institute of Human Nutrition Potsdam-Rehbruecke, Arthur-Scheunert-Allee 114-116, 14558 Nuthetal, Germany
e-mail: behrens@dife.de

W. Meyerhof et al. (eds.), *Sensory and Metabolic Control of Energy Balance*, Results and Problems in Cell Differentiation 52,
DOI 10.1007/978-3-642-14426-4_8, 

balance, whereas sour and bitter warn us against the ingestion of spoiled, unripe, or even toxic foodstuff.

On a cellular level, the various tastants are recognized by taste receptor cells, which are within the oral cavity assembled to taste buds consisting of about 100 cells. Four cell types can be distinguished based on their morphological and physiological characteristic: type-I cells serve a glial-like function, type-II cells express, strictly separated in distinct subsets, taste receptor proteins specific for sweet, umami, and bitter tastants, type-III cells are the only cell type within the taste bud capable of forming synapses with afferent nerve fibers and express sour taste receptors and, finally, type-IV cells are precursor cells able to differentiate

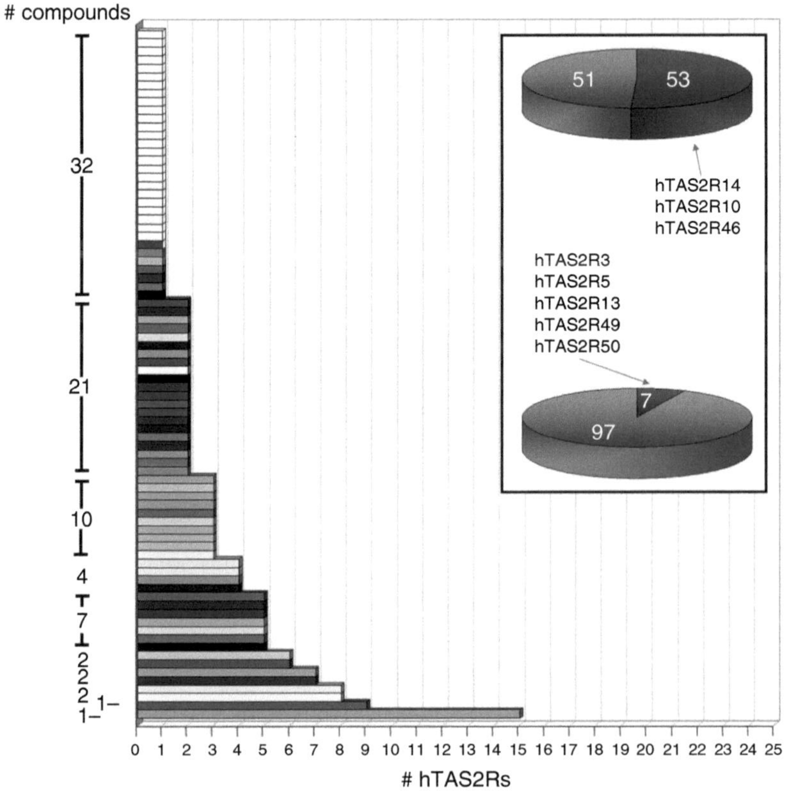


**Fig. 1** *Recognition of bitter compounds by hTAS2Rs.* The recent screening of 25 hTAS2Rs with 104 natural and synthetic compounds (Meyerhof et al. 2010) revealed that each of the 82 identified agonists (*y*-axis) activates a highly variable number of hTAS2Rs (*x*-axis) ranging from 1 to 15 receptors. The inset shows that the apparent breadth of tuning differs considerably among hTAS2Rs. Whereas the three most broadly tuned receptors, hTAS2R14, -R10, and -R46, together detect more than half of all 104 compounds (*upper panel*), the five most narrowly tuned receptors, hTAS2R3, -R5, -R13, -R49, and -R50, recognized merely seven of the tested substances (*lower panel*)

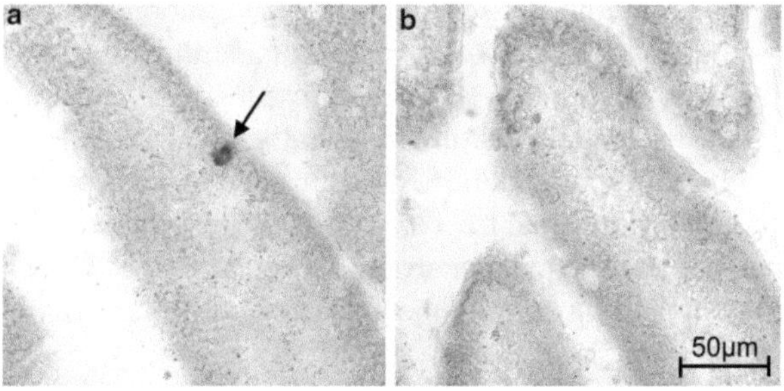


**Fig. 2** *Expression of taste-signaling components in the gastrointestinal tract of mice.* In situ hybridization of duodenal cross-sections of mice with antisense (**a**) and sense (**b**) probes specific for α-gustducin demonstrates the expression of α-gustducin mRNA (*arrow*) in large isolated mucosal cells typical for enteroendocrine cells

throughout an individual's lifetime. Each taste bud has an apical porous region containing the microvilli of the taste receptor cells to establish a direct contact with tastants present in the oral cavity. On the tongue, all taste buds are localized within one of the three types of epithelial specializations, called fungiform (located on the apical tongue surface), foliate (located on both sides of the posterior tongue), and (circum-)vallate papillae (located on the posterior surface of the tongue). As taste receptor cells do not represent neuronal cells, but specialized secondary sensory cells originating from epithelial precursors, the taste information collected by taste buds is transmitted by afferent nerve fibers entering the taste buds from the basolateral side. In total, three cranial nerves convey gustatory information toward the brain: the VII, contacting taste buds of fungiform papillae and soft palate; the IX, providing afferents for foliate and vallate papillae; and the X, innervating pharyngeal and laryngeal taste buds. Within the brain, taste information is collected in a small part of the brain stem, the nucleus tractus solitarius (NTS) and from this first relay station further distributed to cortical structures.

On a molecular level, gustatory stimuli are detected by various taste receptor proteins belonging to structurally different classes of molecules. Whereas ion channels are believed to facilitate the detection of sour (PKD2L1, PKD1L3 (Huang et al. 2006; Ishimaru et al. 2006; LopezJimenez et al. 2006)) and salty tastants (ENaC (Chandrashekar et al. 2010; Kretz et al. 1999; Stahler et al. 2008)), sweet, umami, and bitter compounds activate G-protein-coupled receptors. The products of the three *Tas1r* genes – *Tas1r1*, *Tas1r2*, and *Tas1r3* – form the heteromeric sweet taste receptor composed of TAS1R2 and TAS1R3 (Max et al. 2001; Montmayeur et al. 2001; Nelson et al. 2001) as well as the umami taste receptor, an assembly of TAS1R1 and TAS1R3 subunits (Li et al. 2002; Nelson et al. 2002). The most complex array of receptor proteins is devoted to the detection of bitter compounds. The *Tas2r* gene family (Adler et al. 2000; Chandrashekar et al. 2000; Matsunami et al. 2000) consists of ~30 members in mammals, although considerable fluctuations in gene numbers are observed throughout the vertebrate lineage (Dong et al. 2009).

While already one of the earliest reports on vertebrate bitter taste receptor genes indicated an expression of *Tas2r* genes outside the gustatory system (testis, (Matsunami et al. 2000)), initially *Tas2r* gene expression was believed to be very much restricted to taste-related structures within the oral cavity. However, recently, an increasing number of reports identified *Tas2r* gene expression in other tissues as well. Of these, the gastrointestinal tract as well as the respiratory system received an ever increasing attention. The present article summarizes the current knowledge on *Tas2r* gene expression and function in the gustatory, respiratory, and gastrointestinal systems.

# 2 *Tas2r* Gene Function

## *2.1 Signal Transduction*

Long before the identification of the actual taste receptor molecules, the taste-specific Gα protein, α-gustducin, was discovered (McLaughlin et al. 1992) and subsequently shown to play a crucial role for bitter, sweet (Wong et al. 1996, 1999), and, at least in part, also for umami taste transduction (He et al. 2004). α-Gustducin-positive cells also express the β-subunits Gβ1 and Gβ3 and the γ-subunit Gγ13 (Huang et al. 1999; Rossler et al. 2000) supporting a role of a heterotrimeric G protein with a subunit composition of Gα-gustducin, Gβ1 or Gβ3, Gγ13 for taste transduction. Also the signal transduction components after the initially activated heterotrimeric G protein are well established. Biochemical experiments (Ogura et al. 1997; Rossler et al. 1998) and genetically modified mouse models (Dotson et al. 2005; Zhang et al. 2003) demonstrate the involvement of phospholipase Cβ2 (PLCβ2). Activation of PLCβ2 results in the generation of $IP_3$ causing transient increase in intracellular calcium ions by activation of the ER-membrane-resident type-III $IP_3$ receptor, which was shown to be present in taste cells (Clapp et al. 2001). Finally, transient receptor potential melastatin channel subtype 5 (TRPM5) is activated by the changing intracellular calcium levels causing depolarization of taste receptor cells. Again, the crucial role of TRPM5 for taste signaling was confirmed by genetically modified mouse models (Damak et al. 2006; Talavera et al. 2005; Zhang et al. 2003). Taken together, it is widely accepted that the signaling components briefly described in this paragraph are the main components for taste G-protein-coupled receptor-dependent signal transduction in the gustatory system and therefore represent the canonical signal transduction cascade. Moreover, the above-mentioned signaling molecules are often used as surrogate markers for cells responding to tastants beyond the oral cavity of mammals. However, it should be noted here that a number of observations ranging from residual taste responses in various knock-out models and colocalization studies of taste-related molecules to the presence of alternative signaling components present in subsets of taste cells do not allow to draw conclusions on the exclusiveness of canonical taste transduction components for taste responsiveness in mammals. For a review focusing in more detail on taste signal transduction see Margolskee (2002).

## 2.2 *Pharmacological Characterization of Mammalian Bitter Taste Receptors*

Within the genomes of most mammalian species, 20–40 functional bitter taste receptor genes can be found (Dong et al. 2009). At present, only some mammalian bitter taste receptors have been deorphaned with a clear bias toward human TAS2Rs. Whereas bitter compounds activating 20 of the ~25 human TAS2Rs have been identified (Meyerhof et al. 2010), only 2 of the ~33 mouse Tas2rs (Chandrashekar et al. 2000), a single rat (of ~36 Tas2rs) (Bufe et al. 2002), and one chimpanzee receptor (Wooding et al. 2006) were deorphaned. As many bitter compounds exhibit pronounced pharmacological activities independent of bitter taste receptors, a more detailed analysis of the receptor properties in species other than human would be highly desirable in the future, to allow a better separation between TAS2R-dependent and -independent physiological activities. Focusing on human receptors, it is well documented that they differ considerably in their apparent breadth of tuning. The most broadly tuned receptors are hTAS2R10 (Bufe et al. 2002), hTAS2R14 (Behrens et al. 2004), and hTAS2R46 (Brockhoff et al. 2007). A recent screening of all 25 hTAS2R2s with 104 natural and synthetic bitter compounds (Fig. 1) revealed that each of the three receptors responded roughly to one-third of all substances tested and that their combined activity would be sufficient to detect about half of all bitter compounds (Meyerhof et al. 2010). On the opposite side of the spectrum, another group of receptors exhibits a rather limited breadth of tuning. This group consists of the receptors hTAS2R3 (Meyerhof et al. 2010), -R5 (Meyerhof et al. 2010), -R8 (Pronin et al. 2007), -R13 (Meyerhof et al. 2010), -R49 (Meyerhof et al. 2010), and -R50 (Behrens et al. 2009), which responded to 10 times fewer compounds ($\leq$3 of 104) (Meyerhof et al. 2010). Intermediate promiscuity was observed for the majority of hTAS2Rs responding to 6–20% of the bitter compounds. This group of receptors includes hTAS2R1 (Maehashi et al. 2008), -R4 (Chandrashekar et al. 2000), -R7 (Sainz et al. 2007), -R16 (Bufe et al. 2002), -R38 (Bufe et al. 2005; Kim et al. 2003), -R39 (Meyerhof et al. 2010), -R40 (Intelmann et al. 2009), -R43 (Kuhn et al. 2004; Pronin et al. 2004), -R44 (Kuhn et al. 2004), and -R47 (Pronin et al. 2004). Surprisingly, Meyerhof et al. also demonstrated that the two receptors hTAS2R16 and hTAS2R38 that were considered to be specifically activated by β-D-glucopyranoside and isothiocyanate moieties, respectively, responded to compounds not matching these proposed chemical subclasses. As the two receptors still predominantly detect β-D-glucopyranoside and isothiocyanate containing bitter substances, it appears justified to consider them as specifically tuned receptors.

Another deorphanized receptor with, most likely, intermediate promiscuity is hTAS2R9 (Dotson et al. 2008).

# 3 Respiratory System

Until recently, the mechanism underlying responses of the respiratory system to the application of noxious substances such as pepper was somewhat enigmatic. It was assumed that irritating substances stimulate free nerve endings of the trigeminal nerve, which express various receptors including those responding to hot (capsaicin) and cooling (menthol) agents belonging to the TRP-channel family (Liu and Simon 2000; McKemy et al. 2002). As the trigeminal nerve endings terminate below the epithelial apical tight junction complex (Finger et al. 1990), the way by which the compounds reach the receptors located on these nerve fibers was subject to speculation. The identification of solitary chemosensory cells (SCCs) innervated by trigeminal nerve fibers in the nasal cavity of mice provided an explanation of how the stimulation by especially hydrophilic noxious substances that are not able to penetrate epithelial tight junctions by a paracellular mechanism as speculated for lipophilic compounds may occur. The SCCs express taste-specific signaling components such as α-gustducin, PLCβ2 (Finger et al. 2003), TRPM5 (Gulbransen et al. 2008; Lin et al. 2008), reach the nasal cavity with their apical microvilli, and respond to stimulation with bitter compounds (Finger et al. 2003) and odorous irritants (Lin et al. 2008). The experimental stimulation of nasal SSCs results in respiratory depression in mice indicating a protective mechanism against the inhalation of noxious compounds (Finger et al. 2003; Lin et al. 2008). Further analyses of nasal SCCs revealed that they express bitter taste receptor genes, thus explaining their responsiveness upon application of bitter compounds. Recently, it was shown that a natural source of respiratory irritants may be produced by Gram-negative respiratory pathogens such as *Pseudomonas aeruginosa*. These bacteria produce, among other compounds, acyl-homoserine lactones as signals for their population density (quorum sensing). Indeed, bacterially produced acyl-homoserine lactones, which are chemically related to bitter sesquiterpene lactones, stimulated mouse SSCs via bitter taste-signaling components leading to changes in the respiratory rates of mice treated with those compounds (Tizzano et al. 2010). However, due to the small number of deorphaned mouse bitter taste receptors, it remains to be determined which TAS2R(s) might mediate the observed effect.

Surprisingly, bitter taste receptor expression was recently demonstrated in ciliated cells of human airways (Shah et al. 2009). These cells exhibit motile cilia and have an important function in moving mucus out of the lung in a cell autonomous fashion. By mRNA analyses using human airway epithelia or differentiated cultures thereof, several h*TAS2R* genes along with the signaling components α-gustducin and PLCβ2 were identified. Using antibodies specific for hTAS2R4, -R38, -R43, and -R46, the authors located the bitter taste receptors directly in the ciliary membranes. Strikingly, the ciliary beat frequency in differentiated human airway epithelia was elevated upon stimulation with the bitter compound denatonium (Shah et al. 2009).

## 4 Gastrointestinal System

More than a decade ago, the extraoral expression of the taste-signaling component, α-gustducin (McLaughlin et al. 1992), was demonstrated in the stomach, duodenum (Hofer et al. 1996), and pancreatic ducts (Hofer and Drenckhahn 1998) of rats. Based on their specific morphology, which resemble in many aspects taste receptor cells of the tongue, the gustducin-positive cells were identified as brush cells containing an apical tuft of microvilli (for a review on brush cells, see Sbarbati and Osculati (2005)). More recently, Hass and colleagues detected within the gastric mucosa of mice numerous gustducin-positive cells, of which some were densely clustered in an area between the cardiac and glandular part of the stomach (limiting ridge) (Hass et al. 2007). Utilizing additional marker molecules, the authors of this study demonstrated convincingly that the gustducin-positive cell population in the gastric mucosa consists of at least two distinguishable cell types. One cell type was identified as brush cells, thus confirming previous reports (Hofer et al. 1996; Wu et al. 2002); another cell type is morphologically dissimilar and clustered. Whereas all gustducin-expressing cells are likely to express TRPM5 as well, PLCβ2, although direct double-labeling experiments were not performed, appeared to be expressed mostly in a distinct subset of cells. However, the close association of PLCβ2-positive cells, which also express PGP9.5 and ghrelin, with gustducin-expressing cells suggests a functional relationship of these cell types (Hass et al. 2007).

In a comprehensive study, Bezençon and colleagues detected α-gustducin, PLCβ2, and TRPM5 in stomach, small intestine, and colon of mice (Bezencon et al. 2006). The authors of this study also observed a considerable variance in the overlap of the expression of these molecules on a cellular level. Whereas the majority of cells expressing green-fluorescent protein under the control of the TRPM5 promoter in duodenal villi also express α-gustducin and PLCβ2, a similar degree of colocalization was not observed in other regions of the gut, especially in colon sections PLCβ2-expressing cells appeared to be a rather independent population (Bezencon et al. 2006). The apparent complexity of gustducin-expressing cell types in the gut was confirmed by a recent study in mouse small intestine where three types of gustducin-positive cells were identified expressing gustducin only, gustducin and glucagon-like peptide-1 (GLP-1), or gustducin and 5-HT (Sutherland et al. 2007). The colabeling of gustducin (Fig. 2) and GLP-1 is in good agreement with numerous reports on the role of enteroendocrine cells and cell lines originating from enteroendocrine cells for gastrointestinal nutrient sensing (Chen et al. 2006; Dotson et al. 2008; Dyer et al. 2005; Jang et al. 2007; Jeon et al. 2008; Kokrashvili et al. 2009; Margolskee et al. 2007; Rozengurt et al. 2006; Saitoh et al. 2007; Wu et al. 2002, 2005). These enteroendocrine cells are dispersed within the gut epithelium and secrete numerous hormones, such as GLP-1, GLP-2, PYY (L-cells), GIP (K-cells), and 5-HT (enterochromaffin cells) to name just a few (for a recent review on chemosensing by enteroendocrine cells, see (Sternini et al. 2008).

Concerning the expression of bitter taste receptors in gastrointestinal cells, most studies used RT-PCR experiments to identify TAS2R mRNA in human or rodent gastrointestinal tissue or model cell lines for enteroendocrine cells (Chen et al. 2006; Dotson et al. 2008; Rozengurt et al. 2006; Wu et al. 2002, 2005). At present, only a single cellular colocalization experiment demonstrated the presence of mouse TAS2R138 with chromogranin A, a marker for enteroendocrine cells, in sections of mouse small intestine (Jeon et al. 2008). Intriguingly, the authors of the latter study showed that the transcription factor sterol regulatory element-binding protein-2 (SREBP-2) directly acts on the 5′-upstream region of the *mTas2r138* gene, which contains the putative promoter leading to elevated transcript levels. This explains the observation that phenylthiocarbamide (PTC) stimulation of mTAS2R138-expressing STC-1 cells leading to cholecystokinin (CCK) as well as GLP-1 was enhanced by elevated SREBP-2 levels (Jeon et al. 2008). The PTC-induced secretion of CCK and GLP-1 was dependent on extracellular calcium ions confirming previous experiments demonstrating involvement of L-type voltage-sensitive calcium channels (Chen et al. 2006).

One of the proposed physiological consequences of bitter compounds present in the gastrointestinal tract is a delay in gastric emptying. As bitter compounds represent an aversive taste stimulus inhibiting feeding, Glendinning and colleagues sought to separate oral and gastrointestinal stimuli by intragastric infusion of denatonium solutions in rodent experiments (Glendinning et al. 2008). They demonstrated that 10 mM, but not 2.5 mM, denatonium solution produced robust conditioned taste aversion in rats and that gastric emptying of treated rats occurred at a slower rate. Although the contribution of a dilution effect, which was due to the orally consumed test solution at the same time, was taken into account, the 5 mM effective intragastric concentration of denatonium was higher than the denatonium concentration necessary to condition taste aversion orally (1.25 mM). The same study showed that mice display a similar flavor aversion by gastric infusion of 12 mM denatonium (~6 mM effective intragastric concentration) solution. The duration to induce reduced licking of test solutions in rats after gastric infusion was much longer (>5 min) than the delay observed for oral stimulation (~5 s). If bitter tastants also affect gastric emptying in humans is unclear, since few studies were performed presenting contrasting evidence for (Wicks et al. 2005) or against (Little et al. 2009) an influence of bitter compounds on the slowing of gastric emptying.

The mechanism by which bitter compounds present in the upper gastrointestinal tract influence gastrointestinal function seems to involve vagal afferents as vagal nerve transection below the diaphragm prevents the activation of NTS neurons as visualized by c-fos expression (Hao et al. 2008). It is speculated that activation of vagal nerve fibers, which contain CCK1 and YY2 receptors and terminate close to enteroendocrine cells, involves release of CCK and peptide YY stimulated by bitter compounds (Hao et al. 2008). In a follow-up study, Hao and colleagues identified additional regions within the rat brain responding to intragastric gavage of bitter compounds. They found significantly elevated c-fos expression in parts of the NTS, medulla, parabrachial nucleus, amygdale, and hypothalamus indicating a complex activation pattern, which can be correlated with the observed conditioned taste aversion by bitter compounds (Hao et al. 2009). Within human and rat large intestine, 6-*n*-propyl-2-thiouracil (PROP) evokes anion secretion indicative for an increased transepithelial ion transport (Kaji et al. 2009). The authors correlate the observed activity of PROP with the presence of TAS2Rs detected by RT-PCR. As pathophysiological levels of PGE2 induce inflammation and potentiate PROP-induced effects, it is suggested that noxious substances are flushed out from the colonic lumen faster.

The function of bitter taste receptors in gastrointestinal cells associated with metabolic regulation, such as enteroendocrine cells secreting GLP-1 (L cells) or GIP (K cells), may also explain the observation that an inactive variant of the human bitter taste receptor hTAS2R9 is associated with the occurrence of metabolic phenotypes in an Amish family diabetes study (Dotson et al. 2008).

# 5 Outlook

During recent years, numerous reports indicated nongustatory expression and function of bitter taste receptors. One obvious conclusion from the fact that bitter taste receptors were identified in respiratory and gastrointestinal tissues is that one can no longer consider TAS2Rs as pure taste receptors. However, at present, numerous questions remain to be solved especially in the field of gastrointestinal bitter taste receptor function. (1) What are the cell type(s) expressing bitter taste receptors in vivo? By immunohistochemical methods, in situ hybridization or transgenic techniques the exact nature and localization of TAS2R-containing gut cells has to be established. (2) What is the biological significance of bitter compound-stimulated responses of enteroendocrine cells? As it was shown that enteroendocrine L-cells respond to sweet (Jang et al. 2007; Margolskee et al. 2007), bitter (Chen et al. 2006; Dotson et al. 2008; Rozengurt et al. 2006; Wu et al. 2002), and even all five taste qualities (Saitoh et al. 2007); although the different taste qualities indicate very different properties of the food consumed, it appears difficult to understand what the physiological role of this stereotypical response behavior might be. (3) Are bitter taste receptor molecules involved in the activation of enteroendocrine cells? Most bitter compounds exert especially at high concentrations pronounced pharmacological activity, which does not necessarily involve an interaction with bitter taste receptors. The enormous increase in functionally characterized human bitter taste receptors should allow a pharmacological profiling of established cellular models of GI cells and to correlate the obtained functional data with the expression of h*TAS2R* genes nowadays.

# References

Adler E, Hoon MA, Mueller KL, Chandrashekar J, Ryba NJ, Zuker CS (2000) A novel family of mammalian taste receptors. Cell 100:693–702

Behrens M, Brockhoff A, Kuhn C, Bufe B, Winnig M, Meyerhof W (2004) The human taste receptor hTAS2R14 responds to a variety of different bitter compounds. Biochem Biophys Res Commun 319:479–485

Behrens M, Brockhoff A, Batram C, Kuhn C, Appendino G, Meyerhof W (2009) The human bitter taste receptor hTAS2R50 is activated by the two natural bitter terpenoids andrographolide and amarogentin. J Agric Food Chem 57:9860–9866

Bezencon C, le Coutre J, Damak S (2006) Taste-signaling proteins are coexpressed in solitary intestinal epithelial cells. Chem Senses 32:41–49

Brockhoff A, Behrens M, Massarotti A, Appendino G, Meyerhof W (2007) Broad tuning of the human bitter taste receptor hTAS2R46 to various sesquiterpene lactones, clerodane and labdane diterpenoids, strychnine, and denatonium. J Agric Food Chem 55:6236–6243

Bufe B, Hofmann T, Krautwurst D, Raguse JD, Meyerhof W (2002) The human TAS2R16 receptor mediates bitter taste in response to beta-glucopyranosides. Nat Genet 32:397–401

Bufe B, Breslin PA, Kuhn C, Reed DR, Tharp CD, Slack JP, Kim UK, Drayna D, Meyerhof W (2005) The molecular basis of individual differences in phenylthiocarbamide and propylthiouracil bitterness perception. Curr Biol 15:322–327

Chandrashekar J, Mueller KL, Hoon MA, Adler E, Feng L, Guo W, Zuker CS, Ryba NJ (2000) T2Rs function as bitter taste receptors. Cell 100:703–711

Chandrashekar J, Kuhn C, Oka Y, Yarmolinsky DA, Hummler E, Ryba NJ, Zuker CS (2010) The cells and peripheral representation of sodium taste in mice. Nature 464:297–301

Chen MC, Wu SV, Reeve JR Jr, Rozengurt E (2006) Bitter stimuli induce Ca2+ signaling and CCK release in enteroendocrine STC-1 cells: role of L-type voltage-sensitive Ca2+ channels. Am J Physiol Cell Physiol 291:C726–C739

Clapp TR, Stone LM, Margolskee RF, Kinnamon SC (2001) Immunocytochemical evidence for co-expression of type III IP3 receptor with signaling components of bitter taste transduction. BMC Neurosci 2:6

Damak S, Rong M, Yasumatsu K, Kokrashvili Z, Perez CA, Shigemura N, Yoshida R, Mosinger B Jr, Glendinning JI, Ninomiya Y, Margolskee RF (2006) Trpm5 null mice respond to bitter, sweet, and umami compounds. Chem Senses 31:253–264

Dong D, Jones G, Zhang S (2009) Dynamic evolution of bitter taste receptor genes in vertebrates. BMC Evol Biol 9:12

Dotson CD, Roper SD, Spector AC (2005) PLCbeta2-independent behavioral avoidance of prototypical bitter-tasting ligands. Chem Senses 30:593–600

Dotson CD, Zhang L, Xu H, Shin YK, Vigues S, Ott SH, Elson AE, Choi HJ, Shaw H, Egan JM, Mitchell BD, Li X, Steinle NI, Munger SD (2008) Bitter taste receptors influence glucose homeostasis. PLoS One 3:e3974

Dyer J, Salmon KS, Zibrik L, Shirazi-Beechey SP (2005) Expression of sweet taste receptors of the T1R family in the intestinal tract and enteroendocrine cells. Biochem Soc Trans 33: 302–305

Finger TE, St Jeor VL, Kinnamon JC, Silver WL (1990) Ultrastructure of substance P- and CGRP-immunoreactive nerve fibers in the nasal epithelium of rodents. J Comp Neurol 294:293–305

Finger TE, Bottger B, Hansen A, Anderson KT, Alimohammadi H, Silver WL (2003) Solitary chemoreceptor cells in the nasal cavity serve as sentinels of respiration. Proc Natl Acad Sci U S A 100:8981–8986

Glendinning JI, Yiin YM, Ackroff K, Sclafani A (2008) Intragastric infusion of denatonium conditions flavor aversions and delays gastric emptying in rodents. Physiol Behav 93:757–765

Gulbransen BD, Clapp TR, Finger TE, Kinnamon SC (2008) Nasal solitary chemoreceptor cell responses to bitter and trigeminal stimulants in vitro. J Neurophysiol 99:2929–2937

Hao S, Sternini C, Raybould HE (2008) Role of CCK1 and Y2 receptors in activation of hindbrain neurons induced by intragastric administration of bitter taste receptor ligands. Am J Physiol Regul Integr Comp Physiol 294:R33–R38

Hao S, Dulake M, Espero E, Sternini C, Raybould HE, Rinaman L (2009) Central Fos expression and conditioned flavor avoidance in rats following intragastric administration of bitter taste receptor ligands. Am J Physiol Regul Integr Comp Physiol 296:R528–R536

Hass N, Schwarzenbacher K, Breer H (2007) A cluster of gustducin-expressing cells in the mouse stomach associated with two distinct populations of enteroendocrine cells. Histochem Cell Biol 128:457–471

He W, Yasumatsu K, Varadarajan V, Yamada A, Lem J, Ninomiya Y, Margolskee RF, Damak S (2004) Umami taste responses are mediated by alpha-transducin and alpha-gustducin. J Neurosci 24:7674–7680

Hofer D, Drenckhahn D (1998) Identification of the taste cell G-protein, alpha-gustducin, in brush cells of the rat pancreatic duct system. Histochem Cell Biol 110:303–309

Hofer D, Puschel B, Drenckhahn D (1996) Taste receptor-like cells in the rat gut identified by expression of alpha-gustducin. Proc Natl Acad Sci U S A 93:6631–6634

Huang L, Shanker YG, Dubauskaite J, Zheng JZ, Yan W, Rosenzweig S, Spielman AI, Max M, Margolskee RF (1999) Ggamma13 colocalizes with gustducin in taste receptor cells and mediates IP3 responses to bitter denatonium. Nat Neurosci 2:1055–1062

Huang AL, Chen X, Hoon MA, Chandrashekar J, Guo W, Trankner D, Ryba NJ, Zuker CS (2006) The cells and logic for mammalian sour taste detection. Nature 442:934–938

Intelmann D, Batram C, Kuhn C, Haseleu G, Meyerhof W, Hofmann T (2009) Three TAS2R bitter taste receptors mediate the psychophysical responses to bitter compounds of hops (*Humulus lupulus* L.) and beer. Chem Percept 2:118–132

Ishimaru Y, Inada H, Kubota M, Zhuang H, Tominaga M, Matsunami H (2006) Transient receptor potential family members PKD1L3 and PKD2L1 form a candidate sour taste receptor. Proc Natl Acad Sci U S A 103:12569–12574

Jang HJ, Kokrashvili Z, Theodorakis MJ, Carlson OD, Kim BJ, Zhou J, Kim HH, Xu X, Chan SL, Juhaszova M, Bernier M, Mosinger B, Margolskee RF, Egan JM (2007) Gut-expressed gustducin and taste receptors regulate secretion of glucagon-like peptide-1. Proc Natl Acad Sci U S A 104:15069–15074

Jeon TI, Zhu B, Larson JL, Osborne TF (2008) SREBP-2 regulates gut peptide secretion through intestinal bitter taste receptor signaling in mice. J Clin Invest 118:3693–3700

Kaji I, Karaki S, Fukami Y, Terasaki M, Kuwahara A (2009) Secretory effects of a luminal bitter tastant and expressions of bitter taste receptors, T2Rs, in the human and rat large intestine. Am J Physiol Gastrointest Liver Physiol 296:G971–G981

Kim UK, Jorgenson E, Coon H, Leppert M, Risch N, Drayna D (2003) Positional cloning of the human quantitative trait locus underlying taste sensitivity to phenylthiocarbamide. Science 299:1221–1225

Kokrashvili Z, Mosinger B, Margolskee RF (2009) Taste signaling elements expressed in gut enteroendocrine cells regulate nutrient-responsive secretion of gut hormones. Am J Clin Nutr 90:822S–825S

Kretz O, Barbry P, Bock R, Lindemann B (1999) Differential expression of RNA and protein of the three pore-forming subunits of the amiloride-sensitive epithelial sodium channel in taste buds of the rat. J Histochem Cytochem 47:51–64

Kuhn C, Bufe B, Winnig M, Hofmann T, Frank O, Behrens M, Lewtschenko T, Slack JP, Ward CD, Meyerhof W (2004) Bitter taste receptors for saccharin and acesulfame K. J Neurosci 24:10260–10265

Li X, Staszewski L, Xu H, Durick K, Zoller M, Adler E (2002) Human receptors for sweet and umami taste. Proc Natl Acad Sci U S A 99:4692–4696

Lin W, Ogura T, Margolskee RF, Finger TE, Restrepo D (2008) TRPM5-expressing solitary chemosensory cells respond to odorous irritants. J Neurophysiol 99:1451–1460

Little TJ, Gupta N, Case RM, Thompson DG, McLaughlin JT (2009) Sweetness and bitterness taste of meals per se does not mediate gastric emptying in humans. Am J Physiol Regul Integr Comp Physiol 297:R632–R639

Liu L, Simon SA (2000) Capsaicin, acid and heat-evoked currents in rat trigeminal ganglion neurons: relationship to functional VR1 receptors. Physiol Behav 69:363–378

LopezJimenez ND, Cavenagh MM, Sainz E, Cruz-Ithier MA, Battey JF, Sullivan SL (2006) Two members of the TRPP family of ion channels, Pkd1l3 and Pkd2l1, are co-expressed in a subset of taste receptor cells. J Neurochem 98:68–77

Maehashi K, Matano M, Wang H, Vo LA, Yamamoto Y, Huang L (2008) Bitter peptides activate hTAS2Rs, the human bitter receptors. Biochem Biophys Res Commun 365:851–855

Margolskee RF (2002) Molecular mechanisms of bitter and sweet taste transduction. J Biol Chem 277:1–4

Margolskee RF, Dyer J, Kokrashvili Z, Salmon KS, Ilegems E, Daly K, Maillet EL, Ninomiya Y, Mosinger B, Shirazi-Beechey SP (2007) T1R3 and gustducin in gut sense sugars to regulate expression of Na+-glucose cotransporter 1. Proc Natl Acad Sci U S A 104:15075–15080

Matsunami H, Montmayeur JP, Buck LB (2000) A family of candidate taste receptors in human and mouse. Nature 404:601–604

Max M, Shanker YG, Huang L, Rong M, Liu Z, Campagne F, Weinstein H, Damak S, Margolskee RF (2001) Tas1r3, encoding a new candidate taste receptor, is allelic to the sweet responsiveness locus Sac. Nat Genet 28:58–63

McKemy DD, Neuhausser WM, Julius D (2002) Identification of a cold receptor reveals a general role for TRP channels in thermosensation. Nature 416:52–58

McLaughlin SK, McKinnon PJ, Margolskee RF (1992) Gustducin is a taste-cell-specific G protein closely related to the transducins. Nature 357:563–569

Meyerhof W, Batram C, Kuhn C, Brockhoff A, Chudoba E, Bufe B, Appendino G, Behrens M (2010) The molecular receptive ranges of human TAS2R bitter taste receptors. Chem Senses 35:157–170

Montmayeur JP, Liberles SD, Matsunami H, Buck LB (2001) A candidate taste receptor gene near a sweet taste locus. Nat Neurosci 4:492–498

Nelson G, Hoon MA, Chandrashekar J, Zhang Y, Ryba NJ, Zuker CS (2001) Mammalian sweet taste receptors. Cell 106:381–390

Nelson G, Chandrashekar J, Hoon MA, Feng L, Zhao G, Ryba NJ, Zuker CS (2002) An amino-acid taste receptor. Nature 416:199–202

Ogura T, Mackay-Sim A, Kinnamon SC (1997) Bitter taste transduction of denatonium in the mudpuppy *Necturus maculosus*. J Neurosci 17:3580–3587

Pronin AN, Tang H, Connor J, Keung W (2004) Identification of ligands for two human bitter T2R receptors. Chem Senses 29:583–593

Pronin AN, Xu H, Tang H, Zhang L, Li Q, Li X (2007) Specific alleles of bitter receptor genes influence human sensitivity to the bitterness of aloin and saccharin. Curr Biol 17:1403–1408

Rossler P, Kroner C, Freitag J, Noe J, Breer H (1998) Identification of a phospholipase C beta subtype in rat taste cells. Eur J Cell Biol 77:253–261

Rossler P, Boekhoff I, Tareilus E, Beck S, Breer H, Freitag J (2000) G protein betagamma complexes in circumvallate taste cells involved in bitter transduction. Chem Senses 25:413–421

Rozengurt N, Wu SV, Chen MC, Huang C, Sternini C, Rozengurt E (2006) Colocalization of the alpha-subunit of gustducin with PYY and GLP-1 in L cells of human colon. Am J Physiol Gastrointest Liver Physiol 291:G792–802

Sainz E, Cavenagh MM, Gutierrez J, Battey JF, Northup JK, Sullivan SL (2007) Functional characterization of human bitter taste receptors. Biochem J 403:537–543

Saitoh O, Hirano A, Nishimura Y (2007) Intestinal STC-1 cells respond to five basic taste stimuli. Neuroreport 18:1991–1995

Sbarbati A, Osculati F (2005) A new fate for old cells: brush cells and related elements. J Anat 206:349–358

Shah AS, Ben-Shahar Y, Moninger TO, Kline JN, Welsh MJ (2009) Motile cilia of human airway epithelia are chemosensory. Science 325:1131–1134

Stahler F, Riedel K, Demgensky S, Neumann K, Dunkel A, Taubert A, Raab B, Behrens M, Raguse JD, Hofmann T, Meyerhof W (2008) A role of the epithelial sodium channel in human salt taste transduction? Chem Percept 1:78–90

Sternini C, Anselmi L, Rozengurt E (2008) Enteroendocrine cells: a site of 'taste' in gastrointestinal chemosensing. Curr Opin Endocrinol Diabetes Obes 15:73–78

Sutherland K, Young RL, Cooper NJ, Horowitz M, Blackshaw LA (2007) Phenotypic characterization of taste cells of the mouse small intestine. Am J Physiol Gastrointest Liver Physiol 292: G1420–G1428

Talavera K, Yasumatsu K, Voets T, Droogmans G, Shigemura N, Ninomiya Y, Margolskee RF, Nilius B (2005) Heat activation of TRPM5 underlies thermal sensitivity of sweet taste. Nature 438:1022–1025

Tizzano M, Gulbransen BD, Vandenbeuch A, Clapp TR, Herman JP, Sibhatu HM, Churchill ME, Silver WL, Kinnamon SC, Finger TE (2010) Nasal chemosensory cells use bitter taste signaling to detect irritants and bacterial signals. Proc Natl Acad Sci U S A 107:3210–3215

Wicks D, Wright J, Rayment P, Spiller R (2005) Impact of bitter taste on gastric motility. Eur J Gastroenterol Hepatol 17:961–965

Wong GT, Gannon KS, Margolskee RF (1996) Transduction of bitter and sweet taste by gustducin. Nature 381:796–800

Wong GT, Ruiz-Avila L, Margolskee RF (1999) Directing gene expression to gustducin-positive taste receptor cells. J Neurosci 19:5802–5809

Wooding S, Bufe B, Grassi C, Howard MT, Stone AC, Vazquez M, Dunn DM, Meyerhof W, Weiss RB, Bamshad MJ (2006) Independent evolution of bitter-taste sensitivity in humans and chimpanzees. Nature 440:930–934

Wu SV, Rozengurt N, Yang M, Young SH, Sinnett-Smith J, Rozengurt E (2002) Expression of bitter taste receptors of the T2R family in the gastrointestinal tract and enteroendocrine STC-1 cells. Proc Natl Acad Sci U S A 99:2392–2397

Wu SV, Chen MC, Rozengurt E (2005) Genomic organization, expression, and function of bitter taste receptors (T2R) in mouse and rat. Physiol Genomics 22:139–149

Zhang Y, Hoon MA, Chandrashekar J, Mueller KL, Cook B, Wu D, Zuker CS, Ryba NJ (2003) Coding of sweet, bitter, and umami tastes: different receptor cells sharing similar signaling pathways. Cell 112:293–301

# Reciprocal Modulation of Sweet Taste by Leptin and Endocannabinoids

**Mayu Niki, Masafumi Jyotaki, Ryusuke Yoshida, and Yuzo Ninomiya**

**Abstract** Sweet taste perception is important for animals to detect carbohydrate source of calories and has a critical role in the nutritional status of animals. Recent studies demonstrated that sweet taste responses can be modulated by leptin and endocannabinoids [anandamide (*N*-arachidonoylethanolamine) and 2-arachidonoyl glycerol]. Leptin is an anorexigenic mediator that reduces food intake by acting on hypothalamic receptor, Ob-Rb. Leptin is shown to selectively suppress sweet taste responses in wild-type mice but not in leptin receptor-deficient *db/db* mice. In marked contrast, endocannabinoids are orexigenic mediators that act via $CB_1$ receptors in hypothalamus and limbic forebrain to induce appetite and stimulate food intake. In the peripheral taste system, endocannabinoids also oppose the action of leptin and enhance sweet taste sensitivities in wild-type mice but not in mice genetically lacking $CB_1$ receptors. These findings indicate that leptin and endocannabinoids not only regulate food intake via central nervous systems but also may modulate palatability of foods by altering peripheral sweet taste responses via their cognate receptors.

## 1 Introduction

It is generally known that sensory information of taste is important for evaluating the quality of food components, which is believed to be composed of five basic taste qualities, viz sweet, salty, sour, umami (savory taste), and bitter. Each of these may be responsible for the detection of nutritious and poisonous contents; sweet taste for carbohydrate sources of calories, salty for minerals, umami for protein and amino acids contents, sour for ripeness of fruits and spoiled foods, and bitter for harmful

M. Niki, M. Jyotaki, R. Yoshida, and Y. Ninomiya (✉)
Section of Oral Neuroscience, Kyushu University, Graduate School of Dental Sciences, 3-1-1 Maidashi, Higashi-ku, Fukuoka 812-8582, Japan
e-mail: yuninom@dent.kyushu-u.ac.jp

W. Meyerhof et al. (eds.), *Sensory and Metabolic Control of Energy Balance*, Results and Problems in Cell Differentiation 52,
DOI 10.1007/978-3-642-14426-4_9, 

compounds. The detection of these taste qualities begins with the taste receptors on the apical membrane of taste receptor cells. Recent molecular genetic studies have proposed candidate receptors for the five basic tastes. Sweet, bitter, and umami tastes are mediated by G protein-coupled receptors (GPCRs), such as taste receptor type 1 (T1Rs: sweet and umami) and type 2 (T2Rs: bitter) families. T1R3 combines with T1R2 to form a sweet taste receptor and with T1R1 to form an umami taste receptor (Nelson et al. 2001; Li et al. 2002). Metabotropic glutamate receptors, such as mGluR1 and 4 (brain-expressed and truncated types of the receptors), are also proposed to be umami receptors in addition to T1R1 + T1R3 (Chaudhari et al. 2000; Toyono et al. 2003; San Gabriel et al. 2005). T2Rs are a family of ~25 highly divergent GPCRs; some of them have been identified for their specific bitter ligands (Matsunami et al. 2000; Mueller et al. 2005). Salty and sour tastes are mediated by channel-type receptors; epithelial sodium ion channels (ENaCs) for salty (Heck et al. 1984; Chandrashekar et al. 2010), and acid-sensing ion channels (ASICs) (Ugawa et al. 1998), hyperpolarization-activated cyclic nucleotide-gated potassium channels (HCNs) (Stevens et al. 2001), and polycystic kidney disease 1 L3 and 2 L1 heterodimer (PKD1L3 + PKD2L1) (Ishimaru et al. 2006) for sour (Fig. 1). It has also been shown that each of these receptors is expressed in separate population of taste bud cells, and genetic elimination of taste receptor (or receptor cells) leads to severe loss of sensitivity to a specific taste quality (Nelson et al. 2001; Mueller et al. 2005; Huang et al. 2006; Chandrashekar et al. 2010), suggesting a possibility that different taste bud cells define the different taste modalities and that activation of a single type of taste receptor cells may be sufficient to encode taste quality (Fig. 1). On the other hand, it is also demonstrated that, although about 60% of taste cells are selectively responsive to one of the five basic stimuli, still many taste cells (about 40%) possess sensitivities to multiple taste qualities (Caicedo et al. 2002; Yoshida et al. 2006, 2009; Yoshida and Ninomiya 2010). Furthermore, taste cells with synapses can receive taste signals from the other cells (cell-to-cell communications; Tomchik et al. 2007) and transmit their signals to taste nerve fibers. Therefore, there are still controversial in peripheral taste coding logic.

Recently, it has been shown that taste function can be modulated by hormones or other factors that act on receptors present in the peripheral gustatory system. For example, leptin, an anorexigenic mediator that reduces food intake by acting on hypothalamic receptors (Friedman 2004), selectively suppresses sweet taste responses and these effects may be mediated by leptin receptor, Ob-Rb, expressed in sweet-sensitive receptor cells (Kawai et al. 2000; Shigemura et al. 2004; Nakamura et al. 2008; Sanematsu et al. 2009; Horio et al. 2010; Jyotaki et al. 2010). Glucagon-like peptide-1 (GLP-1), an incretin that influences glucose transport, metabolism, and homeostasis (Rehfeld 1998), normally acts to maintain or enhance sweet taste sensitivity by its paracrine activity (Shin et al. 2008). More recently, we found that sweet-sensitive cells also express receptors for endocannabinoids, orexigenic mediators that induce appetite and stimulate food intake via endocannabinoid ($CB_1$) receptors mainly in hypothalamus (Kirkham et al. 2002). In the peripheral taste system, endocannabinoids also oppose the action of leptin and enhance sweet taste responses in mice (Yoshida et al. 2010). Thus, leptin and endocannabinoids, therefore, not only

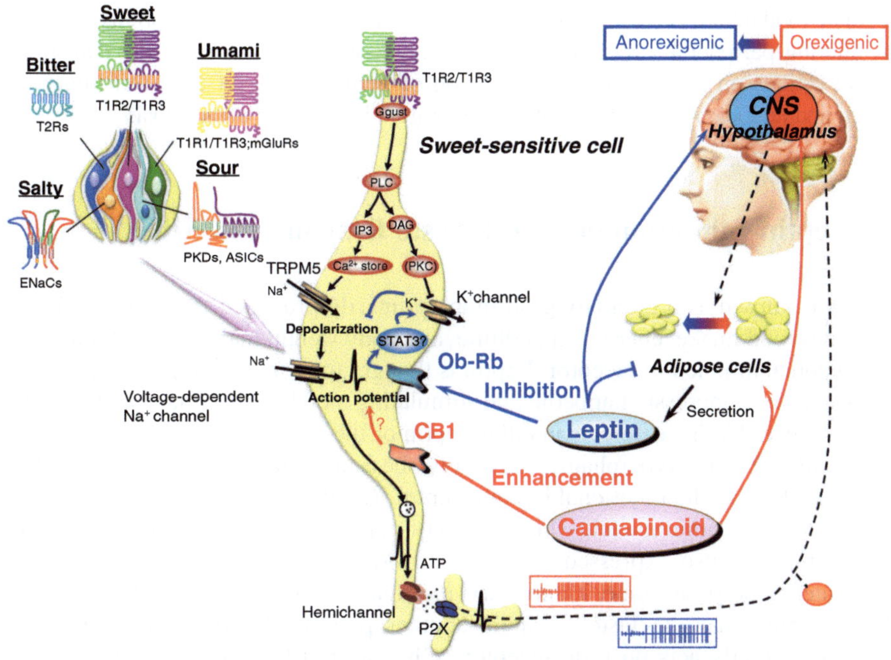


**Fig. 1** Schematic presentations of taste transduction of sweet-sensitive cell expressing leptin receptors(Ob-Rb) and endocannabinoid receptors ($CB_1$) and action of leptin and endocannabinoids on hypothalamus, adipose cells, and sweet-sensitive cell. *Upper left panel*: five basic taste qualities may be encoded by a single type of taste receptor cells in a taste bud which contains 50–150 taste cells. *Middle and right panels*: leptin (anorexigenic factor) and endocannabinoids (orexigenic factor) not only regulate food intake via central nervous systems but also modulate palatability of food by altering peripheral sweet taste sensitivity. The heterodimer of T1R2 + T1R3 was shown to form a sweet taste receptor that can bind a broad array of sweet compounds including external caloric energy sources (Nelson et al. 2001). Binding of sweet compounds to T1R2 + T1R3 leads to activation and dissociation of the subunits of the coupled heterotrimeric G protein, probably Gustducin (Ggust) (Wong et al. 1996) but possibly other G proteins too. The dissociated $\beta\gamma$ subunits of the Ggust activate PLC$\beta$2, which hydrolyzes phosphatidylinositol bisphosphate ($PIP_2$) into diacylglycerol (DAG) and inositol trisphosphate ($IP_3$) (Margolskee 2002). Subsequently, $IP_3$(subscribe) activates the type III $IP_3$ receptor ($IP_3R3$), leading to the release of $Ca^{2+}$ from intracellular stores (Hisatsune et al. 2007). Rapid increases in $[Ca^{2+}]$i open basolaterally located transient receptor potential cation channel, subfamily M, member 5 (TRPM5) channels, leading to the $Na^+$ influx, membrane depolarization, and generation of action potentials (Huang et al. 2002; Medler et al. 2003). The action potential triggers adenosine 5′-triphosphate (ATP) release through hemichannels (Huang et al. 2007; Romanov et al. 2007) and the released ATP activates purinergic P2X receptors in taste fibers which convey sweet taste information to the brain (Finger et al. 2005). TRPM5 channel involved in the transduction of sweet taste acts as a thermosensor (15–35°C). Activation of this channel, by temperature peaking at 35°C near body temperature, leads to increase of the sweet sensitivity of the cell (Talavera et al. 2005). Leptin produced in adipose cells reduces food intake by acting on hypothalamic receptors (Friedman 2004), and selectively suppresses sweet taste responses (Kawai et al. 2000). These effects may be mediated by leptin receptor Ob-Rb expressed in sweet receptor cells. The Ob-Rb may act by STAT3 (signal transducers and activators of transcription) pathway of intracellular signal transduction (Shigemura et al. 2003), and activate

regulate food intake via central nervous systems but also modulate palatability of foods by altering peripheral sweet taste sensitivity. This paper summarized the data from recent studies on the reciprocal modulation of peripheral sweet taste sensitivity by leptin and endocannabinoids and its role in regulating energy homeostasis.

## 2 Leptin Inhibition on Sweet Taste Responses in Mice

Leptin is known as a circulating hormone primarily produced in adipose cells. It regulates food intake, energy expenditure, and body weight mainly via activation of the hypothalamic leptin receptor. Leptin is thought to promote weight loss, at least in rodents, by suppressing appetite and stimulating metabolism (Zhang et al. 1994). Mutant mice that have defects in either leptin or the leptin receptor, such as *ob/ob* and *db/db* mice, are hyperphagic, massively obese, and diabetic (Zhang et al. 1994; Lee et al. 1996). The functional leptin receptor (Ob-Rb) is abundantly expressed in several hypothalamic nuclei, which are major target sites for the hormone. However, Ob-Rb is also expressed in peripheral organs, such as lymph nodes, liver, lung, uterus, adipose tissue, kidney, and pancreas (Flier 2004). Recently, we have demonstrated that the taste organ is another peripheral target for leptin. The hormone directly acts on taste receptor cells via Ob-Rb expressed in these cells and it specifically inhibits peripheral gustatory neural and behavioral responses to sweet substances without affecting responses to sour, salty, and bitter substances in lean mice. By contrast, such selective inhibition of sweet taste responses by leptin was not observed in leptin receptor-deficient *db/db* mice (Kawai et al. 2000; Shigemura et al. 2004; Ninomiya et al. 2002). Instead, the *db/db* mice showed enhanced gustatory neural responses to sweet compounds (Ninomiya et al. 1995, 1998). In lean mice, the strength of suppressive effects by leptin was at most about 30% of control responses, and the effect may saturate when plasma leptin concentration reaches about 15–20 ng/ml (Kawai et al. 2000; Ninomiya et al. 2002). It was also found that potassium outward currents of isolated taste cells in response to depolarizing voltage steps were increased during bath application with leptin to the cells (Kawai et al. 2000). Increase of potassium outward current may lead to reduction of cell excitability and spike frequencies (Fig. 1). In this respect, we have recently found that about a half of sweet-responsive cells showed significant reduction of impulse frequencies in response to sweet stimuli during the application of leptin (10–20 ng/ml) to basolateral membrane of taste cells and their recovery after washout of leptin. (Yoshida et al. unpublished observation).

**Fig. 1** (continued) outward $K^+$ currents, which results in hyperpolarization of taste cells (Kawai et al. 2000). Endocannabinoid that act via $CB_1$ receptors in hypothalamus stimulate food intake (Kirkham et al. 2002) and selectively enhance sweet taste responses (Yoshida et al. 2010). These reciprocal regulations of peripheral sweet taste reception by endocannabinoids and leptin may contribute to their opposing actions on food intake and play an important role in regulating energy homeostasis

## 3 Diurnal Variation of Plasma Leptin Levels and Sweet Taste Recognition Thresholds in Human

Circulating leptin levels are known to have a diurnal variation in both rats and humans (Saladin et al. 1995; Ahima et al. 1996). In humans, plasma leptin levels start rising before noon and peak between 23:00 and 01:00 h, after which the levels decline until morning (Sinha et al. 1996). Leptin, unlike insulin, does not rapidly increase after a meal and does not, by itself, lead to meal termination. However, this does not mean that leptin levels are totally independent of meal timing. Instead, the diurnal variation of leptin levels shows meal-related shifts. For example, when meals were shifted by 6.5 h without changing the light or sleep cycles in humans, the plasma leptin levels were similarly shifted by 5–7 h (Schoeller et al. 1997). The nocturnal rise of leptin does not occur if the subjects are fasted (Boden et al. 1996). Therefore, if leptin acts as a modulator for sweet taste sensitivity, and it shows diurnal variation, then it follows that the threshold for sweet taste may show correlated diurnal variation.

We have recently examined this possibility by measuring recognition thresholds of nonobese subjects (BMI $< 25$) for various taste stimuli and plasma leptin levels at seven time-points (8:00, 9:30, 12:00, 14:00, 17:00, 19:00 and 22:00) during the day under normal meal conditions with three meals, and restricted meal conditions with one or two meals per day (Nakamura et al. 2008). In the normal feeding condition, leptin concentrations started rising before noon and peaked in the night. This rise in leptin occurred later in the two and one meal conditions resulting in a phase shift of diurnal variation. With regard to taste recognition thresholds, similar to plasma leptin levels, significant time-dependent increases in thresholds for sucrose, glucose and saccharin were observed in the normal meal condition (Nakamura et al. 2008). That is, subjects needed higher concentrations of these sweeteners to detect the stimulus quality when they were tested in the evening compared to the morning. There was also a phase shift in one or two meal conditions eliminating the time-dependent changes in sweetener recognition threshold. Diurnal variations in sweetener thresholds were significantly different among three meal conditions. This diurnal variation is sweet-taste selective: it was not observed in thresholds for other taste stimuli (NaCl, citric acid, quinine, and mono-sodium glutamate) (Nakamura et al. 2008). In contrast, the diurnal variations for sweet recognition thresholds became much smaller in the overweight and obese subjects (BMI $> 25$: Shigemura et al. unpublished observation). Plasma leptin levels (at 8:00) of most of the overweight and obese subjects are more than 10 ng/ml, which are close to the saturation level of leptin effect in mice (about 15–20 ng/ml) (Kawai et al. 2000). This suggests that the smaller diurnal variations for sweet recognition thresholds in the over-weight and obese subjects may be due to their higher basal plasma leptin levels.

# 4 Possible Relationships Between Sweet Taste Sensitivity and Postingestive Insulin Responses

In the above-mentioned human study, we also found that blood glucose and plasma insulin levels of nonobese subjects showed meal-related changes with increases evident after each meal in the three different feeding conditions. Moreover, increases in blood glucose after the first meal in the one and two meal conditions were higher than that in the normal, three meal feeding condition. Similar tendency was observed in case of plasma insulin levels. Thus, postingestive rises of blood glucose and insulin levels of individuals after meal were negatively correlated with leptin levels and recognition thresholds for sucrose before meal (Nakamura et al. 2008; Sanematsu et al. 2009; Horio et al. 2010; Jyotaki et al. 2010). This suggests that greater postingestive rises of blood glucose and insulin levels may be associated with lower leptin levels and higher sweet sensitivities before meal.

In relation to this potential linkage, a previous study showed that oral stimulation with sucrose elicits an increase in activities of the pancreatic branch of the vagal nerve in rats, whereas no such response was observed upon stimulation with NaCl (Niijima 1991). The vagal nerve response to sucrose occurs about 5 min after onset of the stimulation and lasts for at least 30 min. Since the so-called cephalic-phase insulin release (CPIR) is known to occur as early as 1–4 min after food ingestion (Teff et al. 1991), the late response observed in the vagal efferent nerve to oral stimulation with sucrose should not relate to the CPIR but may be involved in the factors relate to postingestive insulin release. In addition, recent studies demonstrated that enteroendocrine cells in the gastrointestinal tract express sweet receptors (T1R2 + T1R3) and leptin receptors (Jang et al. 2007; Barrenetxe et al. 2002), and release GLP-1 in response to sugars and nonnutritive sweeteners. This leads to an increase in the expression of $Na^{+}$/glucose cotransporter, SGLT1, followed by increased glucose absorption in enterocytes (Margolskee et al. 2007). Also, sweet receptors (T1R2 + T1R3) and their downstream signaling molecules, gustducin (Ggust) and phospholipase C-β 2 (PLCβ2), are found to be expressed in pancreatic β cells and several artificial sweeteners, such as sucralose, saccharin, and acesulfame-K, are shown to stimulate insulin secretion from the mouse pancreatic islet via sweet receptors on the cell membrane (Nakagawa et al. 2009). Pancreatic β cells are also shown to express leptin receptors (Kieffer et al. 1997). Our recent study demonstrated that enteroendocrine STC-1 cells, like taste cells, responded to sweet compounds and other taste stimuli with rapid increases of intracellular $Ca^{2+}$ concentration, and $Ca^{2+}$ responses to sweet compounds were suppressed by leptin in a concentration-dependent manner (Jyotaki et al. unpublished observation). This suggests a possibility that enteroendocrine cells and pancreatic β cells may also possess comparable sweet sensitivities with diurnal variations parallel with plasma leptin levels and meal-related phase shifts. If this is the case, postingestive rises in glucose and insulin levels may be influenced by sweet sensitivities of both taste and gut and pancreatic β cells (Nakamura et al. 2008).

## 5 Endocannabinoid Enhancement on Gustatory Neural and Behavioral Responses to Sweeteners in Mice

In our previous studies, we found that *db/db* mice with leptin receptor-deficiency showed enhanced responses to sweet compounds (Ninomiya et al. 1995, 1998, 2002). This finding led us to investigate potential effect of leptin on taste responses and resulted in our discovery of inhibitory effect of leptin on sweet taste. However, there still remains unclear why lack of leptin inhibition on sweet taste in *db/db* mice may lead to their enhancement of sweet taste responses (Ninomiya et al. 2002). With regard to our question, Di Marzo and his colleagues reported that defective leptin signaling is associated with elevated hypothalamic levels of endocannabinoids in obese *db/db* and *ob/ob* mice and Zucker rats (Di Marzo et al. 2001). Endocannabinoids, such as anandamide [*N*-arachidonoylethanolamine (AEA)] and 2-arachidonoyl glycerol (2-AG), are known as orexigenic mediators that act via $CB_1$ receptors in hypothalamus and limbic forebrain to induce appetite (Jamshidi and Taylor 2001; Cota et al. 2003) and stimulate food intake (Kirkham et al. 2002). Furthermore, acute leptin treatment of normal rats and *ob/ob* mice reduces AEA and 2-AG in the hypothalamus. $CB_1$ receptor antagonist SR141716 reduces food intake in wild-type mice but not $CB_1$-knockout (KO) mice (Wiley et al. 2005). These findings indicate that endocannabinoids in the hypothalamus appear to be under negative control by leptin and contribute to overeating in the development of obesity. If this is also the case in peripheral taste system, it is possible that enhancement of sweet taste responses observed in *db/db* mice might be due not only to lack of inhibitory effect of leptin but also to the effect of endocannabinoids.

To test this possibility, we first examined gustatory nerve responses of the wild-type mice to various taste stimuli before and after administration of AEA or 2-AG. The responses were obtained from the chorda tympani nerve, innervating the anterior tongue, and the glossopharyngeal nerve, innervating the posterior tongue. The administration of AEA and 2-AG increases responses of both nerves to sweeteners (sucrose, saccharin Na, glucose, and SC45647) in a concentration-dependent manner, whereas no such increase was observed in responses to salty (NaCl), sour (HCl), bitter (quinine), and umami (monosodium L-glutamate) compounds (Yoshida et al. 2010). This suggests that the effect of endocannabinoids is sweet-selective. After i.p. injection of AEA or 2-AG, increased responses to sweet compounds (~150% of control for 500 mM sucrose) were observed at 10–30 min postinjection and then recovered to the control level at 60–120 min postinjection. Consistently, the endocannabinoids selectively increase behavioral lick responses to sucrose–quinine mixtures with similar postinjection time course, whereas no such effect was observed in lick rates for salty, sour, bitter, and umami compounds. By contrast, $CB_1$-KO mice showed no such enhancement of responses to sweet compounds in both gustatory nerve and behavioral response measurements (Yoshida et al. 2010).

# 6 Sweet Enhancing Effect of Endocannabinoids in Taste Cells

We next test the endocannabinoid effect at the taste cell level by comparing responses (action potentials) of single taste cells to sweet compounds before and after the application of AEA or 2-AG to the cells. The results indicate that about 60% of cells (27 of 47) showed enhancement of responses to sweeteners (>120% of control) after the application of 1 μg/ml AEA or 2-AG to the basolateral side of taste cell membrane. The enhancing effects of AEA and 2-AG on sweet responses of taste cells in wild-type mice saturated at ~1 μg/ml. The half-max effective concentration ($EC_{50}$) for enhancing sweet responses of wild-type taste cells by AEA (0.112 μg/ml) was about sixfold greater than that of 2-AG (0.017 μg/ml). The effective concentrations of the endocannabinoids are within the physiological ranges found in various tissues (Sugiura et al. 2002). In $CB_1$-KO mice, sweet responses of taste cells were not affected by 1 μg/ml AEA or 2-AG. A pharmacological blocker of $CB_1$ receptors, AM251, suppressed the sweet enhancing effect of 1 μg/ml 2-AG; however, the $CB_2$ receptor blocker AM630 did not. These data indicate that endocannabinoids act on $CB_1$ receptors to enhance sweet taste responses of taste cells (Yoshida et al. 2010).

Our immunohistochemical study showed that in wild-type mice about 70% of taste cells expressing $CB_1$ receptors coexpressed T1R3; ~60% of taste cells expressing T1R3 also expressed $CB_1$ receptors. In $CB_1$-KO mice, $CB_1$ immunoreactivity in taste cells was absent (Yoshida et al. 2010). In the central nervous system, $CB_1$ receptors are expressed in presynaptic cells and underlie modulation (inhibition) of transmitter release from presynaptic cells (Wilson and Nicoll 2002). In the peripheral taste organ, $CB_1$ immunoreactivity was observed in fewer than 12% of GAD67-expressing taste cells which in mice are thought to be presynaptic cells (Yoshida et al. 2010; DeFazio et al. 2006). GAD67-expressing presynaptic cells are reported to be primarily sensitive to sour taste stimuli (Huang et al. 2008; Yoshida et al. 2009). Endocannabinoids did not affect sour taste responses, indicating that presynaptic cells are not the major target for endocannabinoids in the taste organ. Instead, the majority of taste cells expressing $CB_1$ receptors are sweet-sensitive cells expressing T1R3: endocannabinoids act to enhance sweet taste responses through these taste receptor cells known to lack well-elaborated synapses.

# 7 $CB_1$ Receptor Blocker Abolished Enhanced Sweet Taste Responses in *db*/*db* Mice

We finally examined potential effect of $CB_1$ receptor blocker, AM251, on enhanced sweet taste responses of *db*/*db* mice. The results indicate that the administration of AM251 significantly decreases responses of the chorda tympani nerve to sweeteners (sucrose, saccharin Na, glucose, and SC45647) without affecting responses to salty (NaCl), sour (HCl), bitter (quinine), and umami (monosodium L-glutamate) compounds. This suggests that the effect of AM251 is sweet-selective. After i.p. injection of AM251, decreased responses to sweet compounds (~66% of control for 500 mM sucrose) were observed at 10–30 min postinjection and then recovered to the control level at 60 min postinjection. After administration of AM251, responses to various sweet compounds decrease to the levels similar to those shown by wild-type mice, indicating that enhanced sweet taste responses disappeared in *db/db* mice (Ohkuri et al. unpublished observation). This suggests that enhanced sweet taste responses in *db/db* mice may occur through tonical activation of $CB_1$ by endocannabinoids.

## 8 Functional Significance of Modulation of Peripheral Taste Sensitivities

It has been reported that various peptides, such as leptin (Kawai et al. 2000; Shigemura et al. 2004; Nakamura et al. 2008), cholecystokinin (CCK) (Herness et al. 2002; Shen et al. 2005), vasoactive intestinal peptide (VIP) (Shen et al. 2005), neuropeptide Y (NPY) (Zhao et al. 2005), and GLP-1 (Shin et al. 2008), may be involved in the modulation of peripheral taste sensitivity. Among them, leptin and GLP-1 are shown to be modulators for sweet taste. Leptin specifically suppresses sweet taste responses and these effects may be mediated by leptin receptors (Ob-Rb) on taste cells (Kawai et al. 2000; Ninomiya et al. 2002; Shigemura et al. 2004). GLP-1 signaling increases sweet and sour taste sensitivity and these effects may be mediated by GLP-1 receptors on adjacent intragemmal afferent nerve fibers (Shin et al. 2008). Our above-mentioned study showed that endocannabinoids selectively enhance sweet taste sensitivity via $CB_1$ receptors on the taste cells (Yoshida et al. 2010). Both endocannabinoids and GLP-1 enhance sweet taste but their specificity (sweet vs. sweet and sour) and targets (taste cells vs. afferent fibers) differ, suggesting that these modulators have different roles in modulating sweet taste. Circulating endocannabinoid levels inversely correlate with plasma levels of leptin in healthy human subjects (Monteleone et al. 2005). Both endocannabinoids and leptin affect responses of taste cells via their cognate receptors. Therefore, endocannabinoids and leptin may reciprocally regulate peripheral sweet taste sensitivity.

Previous behavioral studies demonstrated that systemic administration of exogenous cannabinoids or endocannabinoids in rodents causes hyperphagia (Williams and Kirkham 1999) and increases the preference for palatable substances such as sucrose solution or food pellets (Higgs et al. 2003; Jarrett et al. 2005). These effects are mediated by the $CB_1$ receptor: pretreatment with the $CB_1$ receptor antagonist SR141716A inhibited hyperphagia and reduced consumption of both bland and palatable foods (Williams and Kirkham 1999; Higgs et al. 2003; Jarrett et al. 2005). The natural "liking" reactions of rats to sweet compounds were amplified by endogenous cannabinoid signals in nucleus accumbens (Mahler et al. 2007). This suggests that endocannabinoids may be related to hedonic aspects of sweet taste. Our findings that increases in taste cell responses, nerve responses and lick responses to sucrose appear especially at its higher (more palatable) concentrations (Yoshida et al. 2010) are in line with the previous findings obtained from behavioral studies mentioned above. Thus, endocannabinoids not only regulate food intake via central nervous systems but also may modulate palatability of foods by altering peripheral sweet taste responses. Taken together, reciprocal regulation of peripheral sweet taste reception by leptin and endocannabinoids may contribute their opposing actions on food intake and play an important role in regulating energy homeostasis.

**Acknowledgments** This work was supported by Grant-in-Aids 18109013, 1807704 (Y.N.) for Scientific Research from Japan Society for the Promotion of Science.

## References

Ahima RS, Prabakaran D, Mantzoros C, Qu D, Lowell B, Maratos-Flier E, Flier JS (1996) Role of leptin in the neuroendocrine response to fasting. Nature 382:250–252. doi:10.1038/382250a0

Barrenetxe J, Villaro AC, Guembe L, Pascual I, Munoz-Navas M, Barber A, Lostao MP (2002) Distribution of the long leptin receptor isoform in brush border, basolateral membrane, and cytoplasm of enterocytes. Gut 50:797–802. doi:10.1136/gut.50.6.797

Boden G, Chen X, Mozzoli M, Ryan I (1996) Effect of fasting on serum leptin in normal human subjects. J Clin Endocrinol Metab 81:3419–3423

Caicedo A, Kim KN, Roper SD (2002) Individual mouse taste cells respond to multiple chemical stimuli. J Physiol 544:501–509. doi:10.1113/jphysiol.2002.027862

Chandrashekar J, Kuhn C, Oka Y, Yarmolinsky DA, Hummler E, Ryba NJ, Zuker CS (2010) The cells and peripheral representation of sodium taste in mice. Nature 464:297–301. doi:10.1038/nature08783

Chaudhari N, Landin AM, Roper SD (2000) A metabotropic glutamate receptor variant functions as a taste receptor. Nat Neurosci 3:113–119. doi:10.1038/72053

Cota D, Marsicano G, Tschöp M, Grübler Y, Flachskamm C, Schubert M, Auer D, Yassouridis A, Thöne-Reineke C, Ortmann S, Tomassoni F, Cervino C, Nisoli E, Linthorst AC, Pasquali R, Lutz B, Stalla GK, Pagotto U (2003) The endogenous cannabinoid system affects energy balance via central orexigenic drive and peripheral lipogenesis. J Clin Invest 112:423–431. doi:10.1172/JCI17725

DeFazio RA, Dvoryanchikov G, Maruyama Y, Kim JW, Pereira E, Roper SD, Chaudhari N (2006) Separate populations of receptor cells and presynaptic cells in mouse taste buds. J Neurosci 26:3971–3980. doi:10.1523/JNEUROSCI.0515-06.2006

Di Marzo V, Goparaju SK, Wang L, Liu J, Bátkai S, Járai Z, Fezza F, Miura GI, Palmiter RD, Sugiura T, Kunos G (2001) Leptin-regulated endocannabinoids are involved in maintaining food intake. Nature 410:822–825. doi:10.1038/35071088

Finger TE, Danilova V, Barrows J, Bartel DL, Vigers AJ, Stone L, Hellekant G, Kinnamon SC (2005) ATP signaling is crucial for communication from taste buds to gustatory nerves. Science 310:1495–1499. doi:10.1126/science.1118435

Flier JS (2004) Obesity wars: molecular progress confronts an expanding epidemic. Cell 116:337–350. doi:10.1016/S0092-8674(03)01081-X

Friedman JM (2004) Modern science versus the stigma of obesity. Nat Med 10:563–569. doi:10.1038/nm0604-563

Heck GL, Mierson S, DeSimone JA (1984) Salt taste transduction occurs through an amiloride-sensitive sodium transport pathway. Science 223:403–405

Herness S, Zhao FL, Lu SG, Kaya N, Shen T (2002) Expression and physiological actions of cholecystokinin in rat taste receptor cells. J Neurosci 22:10018–10029

Higgs S, Williams CM, Kirkham TC (2003) Cannabinoid influences on palatability: microstructural analysis of sucrose drinking after delta(9)-tetrahydrocannabinol, anandamide, 2-arachidonoyl glycerol and SR141716. Psychopharmacology (Berl) 165:370–377. doi:10.1007/s00213-002-1263-3

Hisatsune C, Yasumatsu K, Takahashi-Iwanaga H, Ogawa N, Kuroda Y, Yoshida R, Ninomiya Y, Mikoshiba K (2007) Abnormal taste perception in mice lacking the type 3 inositol 1,4, 5-trisphosphate receptor. J Biol Chem 282:37225–37231. doi:10.1074/jbc.M705641200

Horio N, Jyotaki M, Yoshida R, Sanematsu K, Shigemura N, Ninomiya Y (2010) New frontiers in gut nutrient sensor research: nutrient sensors in the gastrointestinal tract: modulation of sweet taste sensitivity by leptin. J Pharmacol Sci 112:8–12. doi:10.1254/jphs.09R07FM

Huang L, Rong M, Kozak JA, Preuss AK, Zhang H, Max M, Margolskee RF (2002) A transient receptor potential channel expressed in taste receptor cells. Nat Neurosci 5:1169–1176. doi:10.1038/nn952

Huang AL, Chen X, Hoon MA, Chandrashekar J, Guo W, Tränkner D, Ryba NJ, Zuker CS (2006) The cells and logic for mammalian sour taste detection. Nature 442:934–938. doi:10.1038/nature05084

Huang YJ, Maruyama Y, Dvoryanchikov G, Pereira E, Chaudhari N, Roper SD (2007) The role of pannexin 1 hemichannels in ATP release and cell–cell communication in mouse taste buds. Proc Natl Acad Sci U S A 104:6436–6441. doi:10.1073/pnas.0611280104

Huang YA, Maruyama Y, Stimac R, Roper SD (2008) Presynaptic (Type III) cells in mouse taste buds sense sour (acid) taste. J Physiol 586:2903–2912. doi:10.1113/jphysiol.2008.151233

Ishimaru Y, Inada H, Kubota M, Zhuang H, Tominaga M, Matsunami H (2006) Transient receptor potential family members PKD1L3 and PKD2L1 form a candidate sour taste receptor. Proc Natl Acad Sci U S A 103:12569–12574. doi:10.1073/pnas.0602702103

Jamshidi N, Taylor DA (2001) Anandamide administration into the ventromedial hypothalamus stimulates appetite in rats. Br J Pharmacol 134:1151–1154. doi:10.1038/sj.bjp. 0704379

Jang HJ, Kokrashvili Z, Theodorakis MJ, Carlson OD, Kim BJ, Zhou J, Kim HH, Xu X, Chan SL, Juhaszova M, Bernier M, Mosinger B, Margolskee RF, Egan JM (2007) Gut-expressed gustducin and taste receptors regulate secretion of glucagon-like peptide-1. Proc Natl Acad Sci U S A 104:15069–15074. doi:10.1073/pnas.0706890104

Jarrett MM, Limebeer CL, Parker LA (2005) Effect of delta9-tetrahydrocannabinol on sucrose palatability as measured by the taste reactivity test. Physiol Behav 86:475–479. doi:10.1016/j.physbeh.2005.08.033

Jyotaki M, Shigemura N, Ninomiya Y (2010) Modulation of sweet taste sensitivity by orexigenic and anorexigenic factors. Endocr J 57:467-475. doi:10.1507/endocrj.K10E-095

Kawai K, Sugimoto K, Nakashima K, Miura H, Ninomiya Y (2000) Leptin as a modulator of sweet taste sensitivities in mice. Proc Natl Acad Sci U S A 97:11044–11049

Kieffer TJ, Heller RS, Leech CA, Holz GG, Habener JF (1997) Leptin suppression of insulin secretion by the activation of ATP-sensitive $K^+$ channels in pancreatic β-cells. Diabetes 46:1087–1093

Kirkham TC, Williams CM, Fezza F, Di Marzo V (2002) Endocannabinoid levels in rat limbic forebrain and hypothalamus in relation to fasting, feeding and satiation: stimulation of eating by 2-arachidonoyl glycerol. Br J Pharmacol 136:550–557. doi:10.1038/sj.bjp.0704767

Lee GH, Proenca R, Montez JM, Carroll KM, Darvishzadeh JG, Lee JI, Friedman JM (1996) Abnormal splicing of the leptin receptor in diabetic mice. Nature 379:632–635. doi:10.1038/379632a0

Li X, Staszewski L, Xu H, Durick K, Zoller M, Adler E (2002) Human receptors for sweet and umami taste. Proc Natl Acad Sci U S A 99:4692–4696. doi:10.1073/pnas.072090199

Mahler SV, Smith KS, Berridge KC (2007) Endocannabinoid hedonic hotspot for sensory pleasure: anandamide in nucleus accumbens shell enhances 'liking' of a sweet reward. Neuropsychopharmacology 32:2267–2278. doi:10.1038/sj.npp. 1301376

Margolskee RF (2002) Molecular mechanisms of bitter and sweet taste transduction. J Biol Chem 277:1–4. doi:10.1074/jbc.R100054200

Margolskee RF, Dyer J, Kokrashvili Z, Salmon KSH, Ilegems E, Daly K, Maillet EL, Ninomiya Y, Mosinger B, Shirazi-Beechey SP (2007) T1R3 and gustducin in gut sense sugars to regulate expression of $Na^{+}$-glucose cotranspoter 1. Proc Natl Acad Sci U S A 104:15075–15081. doi:10.1073/pnas.0706678104

Matsunami H, Montmayeur JP, Buck LB (2000) A family of candidate taste receptors in human and mouse. Nature 404:601–604. doi:10.1038/35007072

Medler KF, Margolskee RF, Kinnamon SC (2003) Electrophysiological characterization of voltage-gated currents in defined taste cell types of mice. J Neurosci 23:2608–2617

Monteleone P, Matias I, Martiadis V, De Petrocellis L, Maj M, Di Marzo V (2005) Blood levels of the endocannabinoid anandamide are increased in anorexia nervosa and in binge-eating disorder, but not in bulimia nervosa. Neuropsychopharmacology 30:1216–1221. doi: 10.1038/sj.npp. 1300695

Mueller KL, Hoon MA, Erlenbach I, Chandrashekar J, Zuker CS, Ryba NJ (2005) The receptors and coding logic for bitter taste. Nature 434:225–229. doi:10.1038/nature03352

Nakagawa Y, Nagasawa M, Yamada S, Hara A, Mogami H, Nikolaev VO, Lohse MJ, Shigemura N, Ninomiya Y, Kojima I (2009) Sweet taste receptor expressed in pancreatic beta-cells activates the calcium and cyclic AMP signaling systems and stimulates insulin secretion. PLoS One 4: e5106. doi:10.1371/journal.pone.0005106

Nakamura Y, Sanematsu K, Ohta R, Shirosaki S, Koyano K, Nonaka K, Shigemura N, Ninomiya Y (2008) Diurnal variation of human sweet taste recognition thresholds is correlated with plasma leptin level. Diabetes 57:2661–2665. doi:10.2337/db07-1103

Nelson G, Hoon MA, Chandrashekar J, Zhang Y, Ryba NJ, Zuker CS (2001) Mammalian sweet taste receptors. Cell 106:381–390. doi:10.1016/S0092-8674(01)00451-2

Niijima A (1991) Effects of taste stimulation on the efferent activity of the pancreatic vagus nerve in the rat. Brain Res Bull 26:165–167. doi:10.1016/0361-9230(91)90202-U

Ninomiya Y, Sako N, Imai Y (1995) Enhanced gustatory neural responses to sugars in the diabetic *db/db* mouse. Am J Physiol 269:R930–937

Ninomiya Y, Imoto T, Yatabe A, Kawamura S, Nakashima K, Katsukawa H (1998) Enhanced responses of the chorda tympani nerve to nonsugar sweeteners in the diabetic *db/db* mouse. Am J Physiol 274:R1324–1330

Ninomiya Y, Shigemura N, Yasumatsu K, Ohta R, Sugimoto K, Nakashima K, Lindemann B (2002) Leptin and sweet taste. Vitam Horm 64:221–248

Rehfeld JF (1998) The new biology of gastrointestinal hormones. Physiol Rev 78:1087–1108

Romanov RA, Rogachevskaja OA, Bystrova MF, Jiang P, Margolskee RF, Kolesnikov SS (2007) Afferent neurotransmission mediated by hemichannels in mammalian taste cells. EMBO J 26:657–667. doi:10.1038/sj.emboj.7601526

Saladin R, Vos DP, Guerre-Millo M, Leturque A, Girard J, Staels B, Auwerx J (1995) Transient increase in obese gene expression after food intake or insulin administration. Nature 377:527–529. doi:10.1038/377527a0

San Gabriel A, Uneyama H, Yoshie Y, Torii K (2005) Cloning and characterization of a novel mGluR1 variant from vallate papillae that functions as a receptor for L-glutamate stimuli. Chem Senses 30:i25–i26. doi:10.1093/chemse/bjh095

Sanematsu K, Horio N, Murata Y, Yoshida R, Ohkuri T, Shigemura N, Ninomiya Y (2009) Modulation and transmission of sweet taste information for energy homeostasis. Ann N Y Acad Sci 1170:102–106

Schoeller DA, Cella LK, Sinha MK, Caro JF (1997) Entrainment of the diurnal rhythm of plasma leptin to meal timing. J Clin Invest 100:1882–1887. doi:10.1172/JCI119717

Shen T, Kaya N, Zhao FL, Lu SG, Cao Y, Herness S (2005) Co-expression patterns of the neuropeptides vasoactive intestinal peptide and cholecystokinin with the transduction molecules alpha-gustducin and T1R2 in rat taste receptor cells. Neuroscience 130:229–238. doi:10.1016/j.neuroscience.2004.09.017

Shigemura N, Miura H, Kusakabe Y, Hino A, Ninomiya Y (2003) Expression of leptin receptor (Ob-R) isoforms and signal transducers and activators of transcription (STATs) mRNAs in the mouse taste buds. Arch Histol Cytol 66:253–260. doi:10.1679/aohc.66.253

Shigemura N, Ohta R, Kusakabe Y, Miura H, Hino A, Koyano K, Nakashima K, Ninomiya Y (2004) Leptin modulates behavioral responses to sweet substances by influencing peripheral taste structures. Endocrinology 145:839–843. doi:10.1210/en2003-0602

Shin YK, Martin B, Golden E, Dotson CD, Maudsley S, Kim W, Jang HJ, Mattson MP, Drucker DJ, Egan JM, Munger SD (2008) Modulation of taste sensitivity by GLP-1 signaling. J Neurochem 106:455–463. doi:10.1111/j.1471-4159.2008.05397.x

Sinha MK, Sturis J, Ohannesian J, Magosin S, Stephens T, Heiman ML, Polonsky KS, Caro JF (1996) Ultradian oscillations of leptin secretion in humans. Biochem Biophys Res Commun 228:733–738. doi:10.1006/bbrc.1996.1724

Stevens DR, Seifert R, Bufe B, Müller F, Kremmer E, Gauss R, Meyerhof W, Kaupp UB, Lindemann B (2001) Hyperpolarization-activated channels HCN1 and HCN4 mediate responses to sour stimuli. Nature 413:631–635. doi:10.1038/35098087

Sugiura T, Kobayashi Y, Oka S, Waku K (2002) Biosynthesis and degradation of anandamide and 2-arachidonoylglycerol and their possible physiological significance. Prostaglandins Leukot Essent Fatty Acids 66:173–192. doi:10.1054/plef.2001.0356

Talavera K, Yasumatsu K, Voets T, Droogmans G, Shigemura N, Ninomiya Y, Margolskee RF, Nilius B (2005) Heat activation of TRPM5 underlies thermal sensitivity of sweet taste. Nature 438:1022–1025. doi:10.1038/nature04248

Teff KL, Mattes RD, Engelman K (1991) Cephalic phase insulin release in normal weight males: verification and reliability. Am J Physiol 261:E430–436

Tomchik SM, Berg S, Kim JW, Chaudhari N, Roper SD (2007) Breadth of tuning and taste coding in mammalian taste buds. J Neurosci 27:10840–10848. doi:10.1523/JNEUROSCI.1863-07.2007

Toyono T, Seta Y, Kataoka S, Kawano S, Shigemoto R, Toyoshima K (2003) Expression of metabotropic glutamate receptor group I in rat gustatory papillae. Cell Tissue Res 313:29–35. doi:10.1007/s00441-003-0740-2

Ugawa S, Minami Y, Guo W, Saishin Y, Takatsuji K, Yamamoto T, Tohyama M, Shimada S (1998) Receptor that leaves a sour taste in the mouth. Nature 395:555–556. doi:10.1038/26882

Wiley JL, Burston JJ, Leggett DC, Alekseeva OO, Razdan RK, Mahadevan A, Martin BR (2005) CB1 cannabinoid receptor-mediated modulation of food intake in mice. Br J Pharmacol 145:293–300. doi:10.1038/sj.bjp. 0706157

Williams CM, Kirkham TC (1999) Anandamide induces overeating: Mediation by central cannabinoid (CB1) receptors. Psychopharmacology (Berl) 143:315–317. doi:10.1007/s002130050953

Wilson RI, Nicoll RA (2002) Endocannabinoid signaling in the brain. Science 296:678–682. doi:10.1126/science.1063545

Wong GT, Gannon KS, Margolskee RF (1996) Transduction of bitter and sweet taste by gustducin. Nature 381:796–800. doi:10.1038/381796a0

Yoshida R, Ninomiya Y (2010) New insights into the signal transmission from taste cells to gustatory nerve fibers. Int Rev Cell Mol Biol 279:101–134

Yoshida R, Shigemura N, Sanematsu K, Yasumatsu K, Ishizuka S, Ninomiya Y (2006) Taste responsiveness of fungiform taste cells with action potentials. J Neurophysiol 96:3088–3095. doi:10.1152/jn.00409.2006

Yoshida R, Miyauchi A, Yasuo T, Jyotaki M, Murata Y, Yasumatsu K, Shigemura N, Yanagawa Y, Obata K, Ueno H, Margolskee RF, Ninomiya Y (2009) Discrimination of taste qualities among mouse fungiform taste bud cells. J Physiol 587:4425–4439. doi:10.1113/jphysiol.2009.175075

Yoshida R, Ohkuri T, Jyotaki M, Yasuo T, Horio N, Yasumatsu K, Sanematsu K, Shigemura N, Yamamoto T, Margolskee RF, Ninomiya Y (2010) Endocannabinoids selectively enhance sweet taste. Proc Natl Acad Sci U S A 107:935–939. doi:10.1073/pnas.0912048107

Zhang Y, Proenca R, Maffei M, Barone M, Leopold L, Friedman JM (1994) Positional cloning of the mouse obese gene and its human homologue. Nature 372:425–432. doi:10.1038/372425a0

Zhao FL, Shen T, Kaya N, Lu SG, Cao Y, Herness S (2005) Expression, physiological action, and coexpression patterns of neuropeptide Y in rat taste-bud cells. Proc Natl Acad Sci U S A 102:11100–11105. doi:10.1073/pnas.0501988102

# Roles of Hormones in Taste Signaling

**Yu-Kyong Shin and Josephine M. Egan**

**Abstract** Proper nutrition, avoidance of ingesting substances that are harmful to the whole organism, and maintenance of energy homeostasis are crucial for living organisms. Additionally, mammals possess a sophisticated system to control the types and content of food that we swallow. Gustation is a vital sensory skill for determining which food stuffs to ingest and which to avoid, and for maintaining metabolic homeostasis. It is becoming apparent that there is a strong link between metabolic control and flavor perception. Although the gustatory system critically influences food preference, food intake, and metabolic homeostasis, the mechanisms for modulating taste sensitivity by metabolic hormones are just now being explored. It is likely that hormones produced in the tongue influence the amounts and types of food that we eat: the hormones that we associate with appetite control, glucose homeostasis and satiety, such as glucagon-like peptide-1, cholecystokinin, and neuropeptide Y are also produced locally in taste buds. In this report, we will provide an overview of the peptidergic endocrine hormone factors that are present or are known to have effects within the gustatory system, and we will discuss their roles, where known, in taste signaling.

## 1 Introduction

The intake of proper nutrition, avoidance of ingesting substances that are detrimental, and maintenance of energy homeostasis are crucial for organisms to continue their lives. In mammals, there is a complex system that provides a gateway for controlling the types of food we swallow. Gustation is a vital sensory skill for locating food sources, determining which food stuffs to ingest and maintaining

---

Y.-K. Shin and J.M. Egan (✉)
Diabetes Section/NIA/NIH, 251 Bayview Blvd, Baltimore, MD 21224, USA
e-mail: eganj@grc.nia.nih.gov

W. Meyerhof et al. (eds.), *Sensory and Metabolic Control of Energy Balance*, Results and Problems in Cell Differentiation 52,
DOI 10.1007/978-3-642-14426-4_10, 

**Table 1** Location of hormones and their receptors by cell type within taste buds

| | Type I | Type II | Type III | Type IV | Nerve fibers | References |
|---|---|---|---|---|---|---|
| VIP | | x | | | | Shen et al. (2005) |
| VIP receptor (VPAC1, VPAC2) | | x | | | | Martin et al. (2010) |
| CCK | | x | | | | Shen et al. (2005) |
| CCK receptor (CCK-1) | | x | | | | Herness et al. (2005) |
| NPY | | x | | | | Zhao et al. (2005) |
| NPY receptor (Y1) | | x | | x | | Herness and Zhao (2009) |
| GLP-1 | | x | x | | | Shin et al. (2008) |
| GLP-1 receptor | | | x | | x | Shin et al. (2008) |
| Leptin receptor (Ob-Rb) | | x | | | | Shigemura et al. (2004) |

metabolic homeostasis. It is becoming apparent that there is a strong link between metabolic control and "flavor perception," and endocrine alteration in the taste system is likely to affect food intake, satiety, and general metabolism. Although the gustatory system critically influences food preference, food intake and metabolic homeostasis, the mechanisms for modulating taste sensitivity by metabolic hormones are just now being explored. It is likely that hormones produced in the tongue influence the amounts and types of food that we eat: the hormones that we associate with appetite control, glucose homeostasis, and satiety, such as cholecystokinin (CCK), glucagon-like peptide-1, neuropeptide Y (NPY), are also produced locally in taste buds. In this review, we will provide an overview of the peptidergic endocrine hormone factors that are present or to have effects within the gustatory system and we will discuss their roles, if known, in taste signaling. Table 1 lists the hormones and receptors known to be present in taste buds.

## 2 Hormones That Are Present in the Gustatory System

### *2.1 Vasoactive Intestinal Peptide*

Vasoactive intestinal peptide (VIP) is a 28 amino acid peptide and a member of PACAP/glucagon superfamily that includes secretin, glucagon, and at least 11 other peptides. It was first extracted from the gut as a product of secretin purification and is widely present in the peripheral as well as the central nervous system (Said and Mutt 1970; Fahrenkrug et al. 1979; Ahrén et al. 1980; Lundberg et al. 1980; Lorén et al. 1979). It was defined by its potent smooth muscle relaxant/vasodilatory activity and as a stimulator of secretory activity (Dickson and Finlayson 2009). It serves as a ligand for two G-protein-coupled receptors (GPCRs): VPAC1 and VPAC2 (Martin et al. 2005). These receptors preferentially stimulate adenylate cyclase (AC) and increase intracellular cyclic adenosine monophosphate (cAMP).

Most recently, VIP expression has been identified in the taste cells (TCs) of the rat, hamster, carp, and human (Kusakabe et al. 1998a, b; Herness 1989, 1995; Witt 1995). In mouse TCs, VIP immunoreactivity totally overlaps with PLCβ2 expression (Shen et al. 2005; Fig. 1a); any role, however, of VIP in taste appreciation had not been established prior to 2010.

The majority of VIP-immunoreactive cells also colocalize with α-gustducin, while much fewer VIP-containing cells also express T1R2. α-Gustducin and T1R2 are markers for type II cells, implicating the presence of VIP as having some function in these TCs of the tongue. On the basis of its expression pattern – expression with α-gustducin but little expression with sweet taste receptors – it seemed that VIP might be involved in the transduction of bitter stimuli (Shen et al. 2005). But a recent publication presented data showing that VIP null mice exhibit enhanced taste sensitivity to sweet tastants and they also have heightened sensitivity to bitter stimuli, decreased sensitivity to sour stimuli, and no change in salt perception compared to normal mice. VIP and its receptors (VPAC1, VPAC2) are coexpressed in type II TCs (Fig. 1b and c). Even though VIP null mice had normal gross taste bud morphology, when compared to normal mice, they had significant increases in the number of cells positive for glucagon-like peptide-1 (GLP-1: its role in taste will be discussed later) and leptin receptor expression was

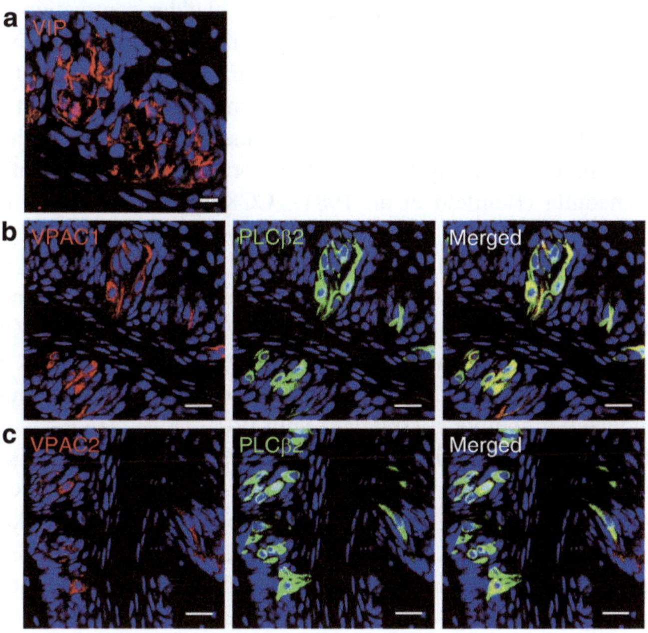


**Fig. 1** *Expression of VIP, and coexpression of VIP receptors (VPAC1/VPAC2) with PLCβ2 in mouse circumvallate papillae (CV).* (**a**) VIP is expressed in taste bud. (**b**) VPAC1 and PLCβ2 are colocalized in a subset of PLCβ2-positive cells. (**c**) VPAC2 and PLCβ2 are colocalized in a subset of PLCβ2-positive cells. Scale bars, 20 μm. *Blue* is TO-PRO-3 nuclear stain

decreased in TCs. The elevated sweet sensitivity of the VIP null mice may stem from the elevated levels of TC GLP-1 (Martin et al. 2010). In a previous study, the authors published that GLP-1's ability to activate GLP-1 receptors in the tongue is strongly involved in the regulation of sweet taste sensitivity in mice (Shin et al. 2008). In addition, Martin et al. found that circulating levels of leptin in VIP null mice were considerably higher than those in normal mice. Therefore, despite the presence of high concentrations of this sweet taste suppressor, the reduction of TC leptin receptor expression was able to facilitate enhanced sweet sensation by preventing inhibitory leptin signaling activity in the tongue. Therefore, significant decrease in leptin receptor expression and elevated expression of GLP-1 may explain sweet hypersensitivity in VIP null mice.

## 2.2 *Cholecystokinin*

CCK is composed of varying numbers of amino acids depending on posttranslational modification of the *CCK* gene product, preprocholecystokinin. So, in reality, CCK is actually a family of hormones identified by the number of amino acids, e.g., CCK58, CCK33, and CCK8. It was first found to be expressed within the enteroendocrine I cells of the proximal small intestine and oral nutrient ingestion, particularly fat, was shown to induce its secretion. Stimulation of gall bladder contraction was the first function ascribed to it, but it plays a role in regulating many gastrointestinal (GI) functions, such as gastric motility and pancreatic enzyme secretion (Buffa et al. 1976; Moran and McHugh 1982). It is also a multifunctional peptide that serves as a neurotransmitter (Larsson and Rehfeld 1979) and it is present in the myenteric nerve plexus, central nervous system, pituitary corticotrophs, C cells of the thyroid, and adrenal medulla (Beinfeld et al. 1981). CCK exerts its actions through two members of the CCK-family of receptors. These receptors, named CCK-1 and CCK-2 (renamed from the original nomenclature of CCK-A and CCK-B respectively) are GPCRs that couple with the inositol trisphosphate (IP3) second messenger system (Foucaud et al. 2008; Rehfeld et al. 2007). Expression of CCK has been reported in the TCs of the rat, hamster, frog, and human (Herness et al. 2002; Kusakabe et al. 1998a, b; Fig. 2a). We have also confirmed CCK immunoreactivity in rat tongue where it completely overlaps with $IP_3R_3$ (Fig. 2a).

More that 50% of CCK-positive TCs also express α-gustducin, an essential component of the bitter pathway; however, very few CCK-containing TCs coexpress T1R2, a sweet taste receptor. Therefore, it was thought that CCK is likely involved in bitter transduction (Lu et al. 2003; Shen et al. 2005). Following this hypothesis, CCK-responsive TCs were characterized for sensitivity to two bitter stimuli, quinine or caffeine, using the calcium-sensitive dye fura-2 as a tool to study intracellular calcium. The investigators found that the same TCs that responded to CCK by increasing intracellular calcium, also responded to quinine and caffeine. The investigators were careful to show that quinine-induced elevations of intracellular calcium were not due to endogenous fluorescence of the quinine molecule. Additionally, when

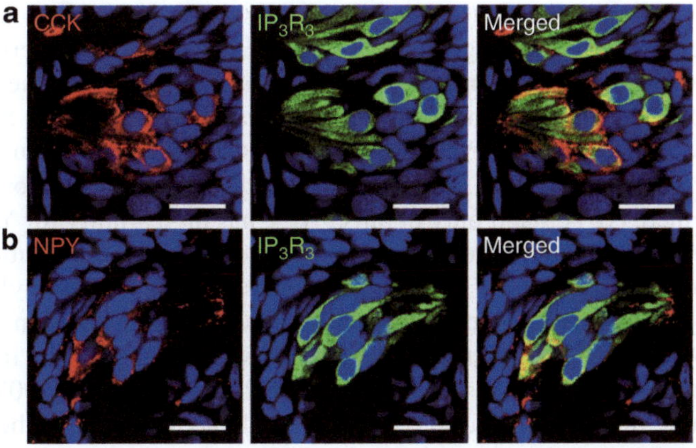


**Fig. 2** *Expression of CCK and NPY in rat circumvallate papillae (CV)*. (**a**) CCK and $IP_3R_3$ are colocalized in a subset of $IP_3R_3$-positive cells in rats. (**b**) NPY and $IP_3R_3$ are colocalized in a subset of $IP_3R_3$-positive cells in rats. Scale bars, 20 μm. *Blue* is TO-PRO-3 nuclear stain

extracellular calcium was depleted, intracellular calcium rose in response to CCK stimulation. On the other hand, when thapsigargin, which depletes intracellular calcium stores, was added to the cells before CCK stimulation, there were minimal increases noted in intracellular calcium. These studies suggest that CCK operates through the known CCK-1 receptor in the DAG/IP3-dependent manner. Most interestingly, about 60–70% of the TCs that responded to CCK also responded to cholinergic stimulation (Lu et al. 2003). CCK was found to increase excitability of TCs by inhibiting outward- and inward-rectifying potassium currents, illustrating that CCK maintains the TCs in a depolarized state for a longer period of time (Herness et al. 2002). For practical purposes, this means that cells that presumably release neurotransmitter during an action potential should have enhanced signaling of the transmitter if CCK were also released from the same cell, which would be expected to accentuate bitter taste. CCK works in an autocrine fashion because there is complete overlap of CCK and CCK-1 receptor-expressing cells, meaning that it's the same cells that release CCK that then has enhanced downstream signaling of the neurotransmitter (Herness et al. 2005). Other modalities besides sweet have not been tested to the same extent. On the basis of pharmacological tools, CCK-2 is not expressed on TCs (Herness et al. 2005).

## 2.3 Neuropeptide Y

NPY, a 36 amino acid peptide, is one of the most abundant neuropeptides in the central nervous system, where it is widely distributed. It is a member of the Y family of peptides, the other family members being pancreatic polypeptide and peptide YY.

It is a potent orexigenic factor (Kalra and Kalra 2004). In addition, it plays a role in a very wide range of physiological processes, including anxiety, energy balance, feeding, vasoconstriction, immune function, reproduction, and heart disease (Gehlert 2004; McDermott and Bell 2007; Pedrazzini et al. 2003; Sperk et al. 2007; Wheway et al. 2007). NPY receptors are GPCRs and at least four Y subtypes have been cloned; Y1, Y2, Y4, and Y5 (Wraith et al. 2000). An additional receptor subtype, termed Y6, is found only in mice. NPY has high affinity for Y1, Y2, and Y5, and the NPY modulatory effect in TCs is mediated by the Y1 subtype (Zhao et al. 2005). Activation of the Y receptors causes inhibition of adenylyl cyclase, mobilization of intracellular calcium via IP3 production, activation of inward-rectifying potassium currents (the opposite to CCK-1 activation), and inhibition of potassium and calcium currents (Pedrazzini et al. 2003; Sperk et al. 2007; Sun et al. 1998; Sun and Miller 1999). In TCs, NPY is found in type II cells of the taste buds (Fig. 2b) where it completely overlaps with CCK- and VIP-expressing cells (Zhao et al. 2005), and, as expected, it completely overlaps with $IP_3R_3$ immunoreactivity (Fig. 2b). Thus, there are some TCs coexpressing CCK, NPY, and CCK-1 receptors. NPY, however, exerts inhibitory actions on single TCs that are opposite to those exerted by CCK (Zhao et al. 2005) and it acts in a paracrine fashion because, unlike CCK and its receptor, NPY and the Y1 receptor subtype are expressed in separate TCs. Using whole cell patch clamp analysis of inward-rectifying potassium channels, the magnitude of $K_{IR}$ currents were significantly enhanced by NPY. In the same patch clamp paradigm, voltage-dependent sodium channels and voltage-dependent outward potassium currents were not altered by NPY. Because $K_{IR}$ is involved in maintaining the resting membrane potential, enhancing the current would cause hyperpolarization, meaning that the cells would be less responsive to action potential-generating stimuli. BIBP3266, a Y1 antagonist, caused a highly significantly reduced response to NPY and the number of NPY-responsive TCs dropped dramatically, implying that Y1 is the receptor responsible for NPY effects in TCs. Y1 coexpresses with T1R3, a sweet receptor essential for sensing sweet taste. CCK and NPY, when they are released together with neurotransmitter from cells during ligand activation of bitter receptors, may work in concert to both upregulate the excitation of bitter-sensitive taste receptor expressing cells while concurrently inhibiting neighboring sweet-sensitive cells (For a more complete review of CCK and NPY in TCs, see Herness and Zhao 2009).

## *2.4 Proglucagon Fragments*

The proglucagon gene encodes glucagon and two glucagon-like peptides that have approximately 50% amino acid homology to glucagon; these are designated GLP-1 and glucagon-like peptide-2 (GLP-2). Glucagon, GLP-1 and GLP-2 are posttranslational cleavage products of proglucagon resulting from prohormone convertase

PC2 and PC1/3 enzymes more usually in islet α cells and enteroendocrine L cells, respectively (Fig. 3a).

Alpha cells are one of the five cells types in islet of Langerhans and enteroendocrine L cells are scattered among the enterocytes and many other enteroendocrine cells throughout the small bowel and ascending colon. However, GLP-1 and GLP-2 are also found in certain brain areas (Doyle and Egan 2007; Jang et al. 2007). In the fasting state, circulating glucagon levels gradually increase, especially in the portal blood circulation, the ultimate goal of glucagon being to protect the brain from neuroglucopenia. Activation of specific glucagon receptors, which are G-protein-coupled, on hepatocytes causes increased intracellular cAMP levels, leading to hepatic glucose production from both glycogenolysis and gluconeogenesis from gluconeogenic precursors (Williams textbook of Endocrinology).

In humans, GLP-1 exists in multiple forms. The majority (at least 80%) of circulating biologically active GLP-1 in humans is the COOH-terminally amidated form, GLP-1 (7-36) amide, with lesser amounts of the minor glycine extended form, GLP-1 (7-37), also detectable (Orskov et al. 1986). GLP-1 is one of two incretins,

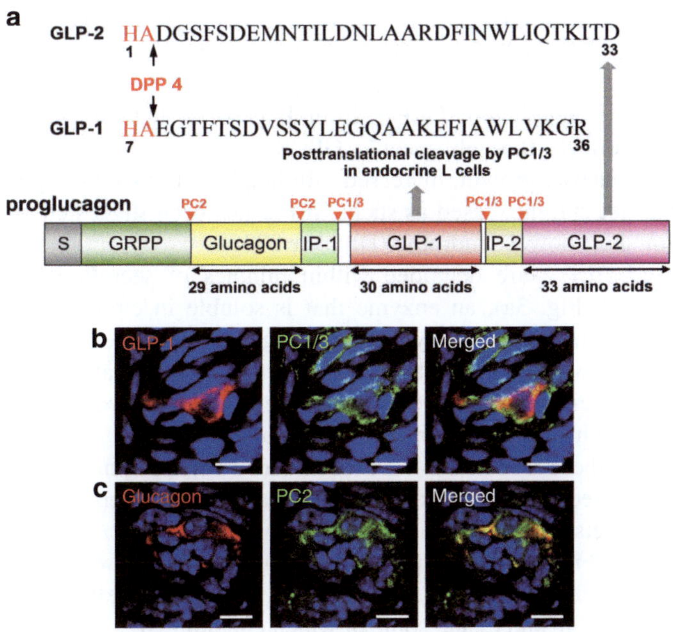


**Fig. 3** *Schematic representation of proglucagon, its cleavage sites and expression of prohormone convertases in mice circumvallate papillae (CV).* (**a**) GLP-1 and GLP-2 are posttranslational cleavage products of the proglucagon gene in enteroendocrine L cells and GLP-1 (7-36) amide is the major form of circulating biologically active GLP-1 in humans. GLP-1 and GLP-2 are rendered biologically inactive by the cleavage of their first two amino acids by dipeptidyl peptidase 4 (DPP4). Glucagon is produced primarily in pancreatic α cells. (**b**) GLP-1 and PC1/3 are colocalized in mice CV. (**c**) Glucagon and PC2 are colocalized in mice CV. Scale bars, 20 μm. *Blue* is TO-PRO-3 nuclear stain

the other being glucose-dependent insulinotropic peptide (GIP). By definition, incretins are hormones that are produced in the gut and secreted into the blood stream in response to food: they enhance insulin secretion (insulinotropism) in a glucose-dependent manner (Creutzfeldt 1979). GLP-1's insulinotropic effects are due to its activation of a specific G-protein-coupled receptor (GLP-1R) that is coupled to increases in intracellular cAMP and $Ca^{2+}$ levels in β cells. It also inhibits gastric emptying, decreases food intake (Willms et al. 1996), inhibits glucagon secretion (Komatsu et al. 1989), and slows the rate of endogenous glucose production (Prigeon et al. 2003), all of which would lead to tight regulation of blood glucose. Additionally, it protects β cells from apoptosis (Farilla et al. 2002) and increases β-cell proliferation by upregulation of a specific β-cell transcription factor pancreatic duodenal homeobox-1 protein (PDX-1) (Perfetti and Merkel 2000; Stoffers et al. 2000), which augments insulin gene transcription leading to increases in insulin production, and upregulates glucose transporter2 (GLUT2) and glucokinase (Wang et al. 1999). Defective secretion of GLP-1 (or GIP) is not considered a cause of diabetes because its secretion after food intake is not decreased in newly diagnosed type II diabetes (T2DM). Continuous GLP-1 treatment in T2DM can normalize blood glucose, improve β-cell function, and restore first-phase insulin secretion and "glucose competence" to β cells (Holz et al. 1993; Zander et al. 2002); hence, GLP-1 analogs and GLP-1R agonists are now treatments for type II diabetes.

GLP-2 is cosecreted with GLP-1 from L cells of the gut. Like GLP-1, it transduces its effects through specific GPCRs. Pharmacological levels lead to increased gut mucosal growth, increased villi height, and increased glucose transport. Though not currently used as such, GLP-2 has been suggested as a possible treatment for short-bowel syndrome (Drucker 2005).

GLP-1 and GLP-2 are degraded within minutes of secretion by dipeptidyl peptidase (DPP4: Fig. 3a), an enzyme that is soluble in circulation and is also membrane-bound in many cells types, including endothelial cells of blood vessels and lymphocytes, where it is also known as CD26. This inactivates both peptides to the extent that their respective receptors are no longer activated by the loss of histadine and alanine on the *N*-terminus of the peptides.

Recently, another role of GLP-1 has been added because functional GLP-1 that activates GLP-1 receptors was extracted from taste buds, where it is expressed in two distinct subsets of TCs that, as expected, also express PC1/3 (Shin et al. 2008; Feng et al. 2008; Fig. 3b). It is expressed in a subset of type II cells that coexpress T1R3 and α-gustducin (Fig. 4a), and a subset of type III cells, some of which contain 5-HT (Fig. 4b and c) and some of which contain PGP 9.5 (Shin et al. 2008).

Taste buds, like brain, are devoid of DPP 4, so the GLP-1 concentration produced in TCs should be high, giving sufficient concentration to activate GLP-1 receptors (Shin et al. 2008). Mice lacking GLP-1 receptors have significantly reduced sweet taste sensitivity to sucrose and sucralose, and they also have a significant hypersensitivity to umami taste (Martin et al. 2009). As regards bitter taste, GLP-1 receptor null mice are fairly similar to control mice (Shin et al. 2008). This work demonstrates that GLP-1 in TCs plays a role in modulating taste

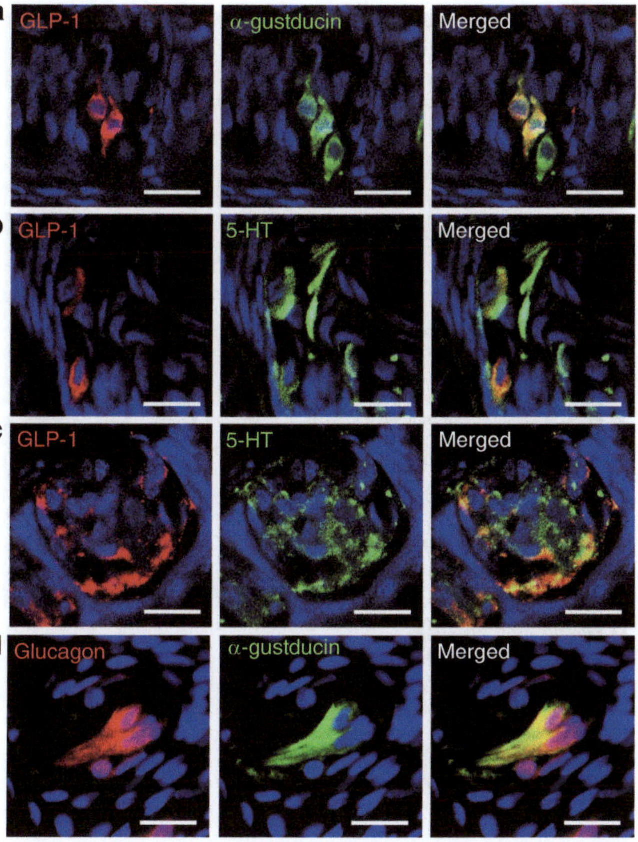


**Fig. 4** *Expression of proglucagon fragments in taste buds in circumvallate papillae.* (**a**) GLP-1 and α-gustducin is colocalized in a subset of α-gustducin-positive cells in rats. (**b**) GLP-1 and 5-HT is colocalized in a subset of 5-HT-positive cells in rats. (**c**) GLP-1 and 5-HT is colocalized in a subset of 5-HT-positive cells in monkey. (**d**) Glucagon and α-gustducin is colocalized in a subset of α-gustducin-positive cells in monkey. Scale bars, 20 μm. *Blue* is TO-PRO-3 nuclear stain

sensitivity through its own receptor, which is present on intragemmal nerve fibers and some PGP 9.5-positive TCs (Fig. 5a and b).

Several studies have shown that many taste-related molecules are found in stomach, intestine, and pancreatic ducts. In addition, some enteroendocrine cells in small intestine (e.g., enteroendocrine L cells) secrete hormones in a T1R3- and α-gustducin-dependent manner in response of sugars (Jang et al. 2007). Our observation that a subset of TCs expressing T1R3 and α-gustducin as well as PLCβ2 and transient receptor potential M5; (Chandrashekar et al. 2006) also express and secrete GLP-1 suggests that they share other molecular mechanisms and/or physiological roles.

While glucagon immunoreactivity seems to be present in type II cells where it colocalizes with α-gustducin (along with PC2, Figs. 4d and 3c) and GLP-2 receptor

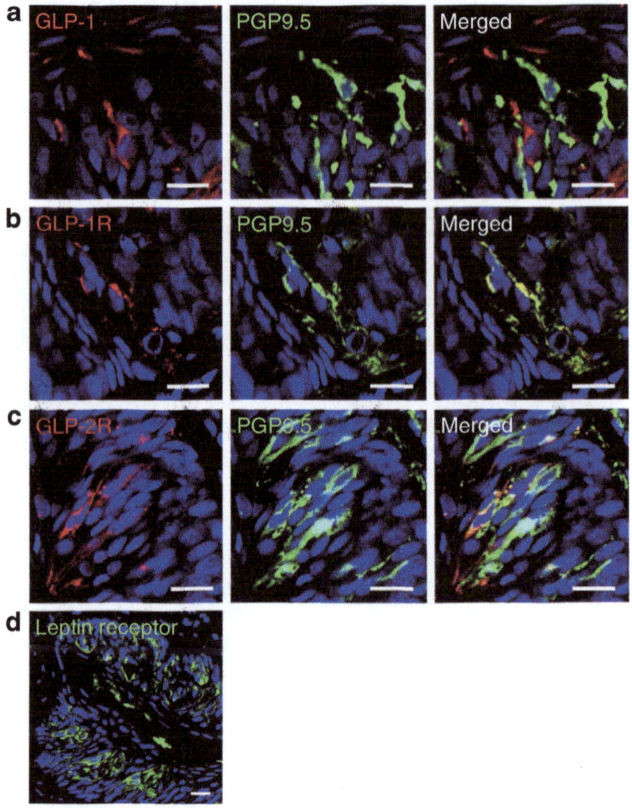


**Fig. 5** *Expression of GLP-1, GLP-1 and GLP-2 receptors, and leptin receptor in rat circumvallate papillae.* (**a**) GLP-1 is expressed in type II and III cells, some of which also stain for PGP 9.5 (see Fig. 3 for additional cell-specific markers). (**b**) GLP-1R and PGP 9.5 are colocalized on nerve fibers and cells. (**c**) GLP-2R and PGP 9.5 are colocalized on cells. (**d**) Leptin receptor is expressed in taste buds in mice. Scale bars, 20 μm. *Blue* is TO-PRO-3

(Fig. 5c), both also found in TCs, any function in taste that they may have has not yet been described in the literature.

## *2.5 Leptin*

Leptin, 167 amino acid product of the obese (*ob*) gene, is a hormone primarily produced in adipocytes. In addition to white adipose tissue – the major source of leptin – it is also produced by brown adipose tissue, lymphocytes, placenta, ovaries, skeletal muscle, stomach (lower part of fundic glands), mammary epithelial cells, bone marrow, pituitary, and liver (Margetic et al. 2002). It plays a role in regulating food intake, energy expenditure, and adipose tissue mass. Its key target organ is the brain where its actions are mediated through leptin receptors in hypothalamic nuclei but leptin receptors are also present in other brain areas and some peripheral organs, including pancreatic islets (Mercer et al. 1996; Hoggard et al. 1997). Leptin acts by binding to a specific "obese receptor" (Ob-R). Several isoforms exist, which are generated as splice variants of one gene, and differ mainly in the length of the cytoplasmic domain (Takaya et al. 1996; Bjørbaek et al. 1998; Yamashita et al. 1998). The leptin receptor is encoded by the *db* gene (Lee et al. 1996). The *db/db* mouse has a point mutation of the *db* gene which leads to abnormal splicing of the coding region. The resulting absence of Ob-Rb, the longer form of Ob-Rs, causes leptin insensitivity, resulting in the obese, diabetic phenotype (Lee et al. 1996). Even before the discovery of leptin, studies of the genetics of taste sensitivity explored a possible role of the *db* gene on sweet taste responses (Sako et al. 1996; Ninomiya et al. 1995, 1998). It was found that diabetic *db/db* mice have greater gustatory neural sensitivities and higher behavioral preferences for various sweet substances than lean control mice. After the discovery of leptin and the finding of receptor defects in Ob-R in *db/db* mice, possible effect of leptin on taste responses were examined in detail. It was found that a subset of TCs is affected by peripheral leptin and that they express Ob-Rb (Fig. 5d), which indicates that TCs are an additional peripheral site of leptin action (Kawai et al. 2000). In situ hybridization (ISH) and mRNA analysis revealed that Ob-Rb mRNAs were expressed in both fungiform and circumvallate taste buds, although no such expression was evident in *db/db* mice.

Ob-Rb is reported to act via the STAT (signal transducers and activators of transcription) pathway of intracellular signal transduction, which is a class of transcription factors having seven members (STAT1, 2, 3, 4, 5a, 5b, and 6) (Frühbeck 2006). By *in vivo* and *in vitro* studies, leptin has been shown to activate JAK2, STAT1, STAT3, STAT5, and STAT6 (Rosenblum et al. 1996; Goïot et al. 2001), whereas *in vivo* studies of the mouse and rat hypothalamus, it specifically activated STAT3 without affecting other components (Vaisse et al. 1996; McCowen et al. 1998; Goïot et al. 2001). In taste papillae, mRNA of STAT members (1–6) was found and the amount of the STAT3 mRNA was higher than that of the other members by mRNA ISH analysis. But, it is still not proven conclusively that STAT3 is coexpressed with Ob-Rb in particular TCs.

Similar to leptin-mediated effects in pancreatic β cells (Harvey and Ashford 1998) and hypothalamic neurons (Spanswick et al. 1997), leptin has been shown to increase $K^+$ conductance of TCs, which, like NPY, results in hyperpolarization and reduction of cell excitability. It has been shown that both nutritive and nonnutritive sweeteners can activate separate signaling transduction cascades, one that involves cAMP and the other that involves inositol triphosphate (IP3), in the same sweet-sensing cell in rat circumvallate taste buds (Bernhardt et al. 1996). A functional role for leptin has been established using lean and *db/db* mice in that leptin plays a role in modulating sweet taste perception. The obese *db/db* mice display enhanced neural responses and elevated behavioral preferences to sweet stimuli (Kawai et al. 2000). Administration of leptin to lean mice suppressed the responses of the peripheral taste nerves to sweet substances, but had no effect on the other taste modalities. Whole-cell patch-clamp recordings performed on TCs have shown that leptin can activate outward $K^+$ currents, which results in a hyperpolarization of the TCs. The *db/db* mice showed no hyperpolarization or leptin suppression, which suggests that leptin acts as a negative modulator for sweet taste (Kawai et al. 2000; Shigemura et al. 2003, 2004). This suppression of sweet taste perception by leptin could subsequently play a role in determining food/calorie intake.

## 2.6 Galanin

Galanin, a neuropeptide with 29 amino acids (30 amino acids in humans), was originally isolated from porcine small intestine (Tatemoto et al. 1983). It is widely distributed in the central and peripheral nervous systems (Branchek et al. 2000; Waters and Krause 2000). Galanin is engaged in the regulation of processes such as food intake (Koegler and Ritter 1998), memory, neuroendocrine function (Mitchell et al. 1999), gut secretion, and motility (Wang et al. 1998a). Galanin mediates its effects through the activation of at least three G-protein-coupled receptor subtypes: galanin receptor (GalR)1, GalR2, GalR3 (Branchek et al. 1998, 2000; Iismaa and Shine 1999). These receptors show distinct distribution patterns and activate different second messenger pathways, depending on the cell type (Chen et al. 1992; Karelson et al. 1995; Wang et al. 1998b; Branchek et al. 2000).

Galanin expression has been observed in the sensory epithelia in the developing ear, eye, and nose (Xu et al. 1996). Later, the expression of galanin has been described for taste bud cells. RT-PCR and ISH experiments showed that mRNA of galanin and GalR2 are detected in rat TCs. Double-label studies uncovered that galanin is present in a subset of α-gustducin-, NCAM-, and PLCβ2-positive TCs, indicating that galanin-expressing TCs are type II and type III cells.

Several studies found that galanin in hypothalamic paraventricular nucleus (PVN) has a role in stimulating the consumption of ethanol and intake of a high-fat diet (Karatayev et al. 2009, 2010). Galanin-overexpressing mice using human dopamine β-hydroxylase promoter showed an increase in ethanol intake and ethanol preference, and no change in consumption of sucrose or quinine solutions in preference tests compared with control mice. In addition, GALOE mice consumed 55% more fat-rich diet during a 2-h test period (Karatayev et al. 2009). Then same group studied galanin knockout (GALKO) mice. The results revealed that GALKO mice had a decrease in ethanol intake and preference, decrease in acute intake of a fat-rich diet, no difference in consumption of sucrose or quinine solutions in preference tests, and a total loss of GAL mRNA in the PVN (Karatayev et al. 2010). These results provide strong support for a physiological role of galanin in stimulating the consumption of ethanol, as well as a fat-rich diet.

# 3 Hormones That Are Not Expressed in Taste Buds but Modify Taste Preference

## 3.1 *Vasopressin*

Vasopressin is nine amino acid peptide hormone. It is derived from a preprohormone precursor that is synthesized in the hypothalamus and transported to the posterior pituitary. One site of action is in the kidneys where it is necessary for the conservation of water by concentrating the urine and reducing urine volume, and it is based on this effect that it received its other name, antidiuretic hormone (ADH). It is also involved with regulation of cardiovascular processes because it is involved in pressure and osmolality regulation, which explains its second name of vasopressin and temperature regulation (Chase and Aurbach 1968; Cooper et al. 1979). Vasopressin exerts its effects through receptors that have been divided two broad classes, the V1 and V2 receptors. The receptors differ in their distribution and associated second messenger systems (Jard 1988). V1 receptors are located on blood vessels where they mediate vasopressin's cardiovascular effects, and V2 are located on renal collecting tubules where they control water retention.

Brain release and pituitary release of vasopressin often parallel one another. For example, systemic osmotic and hypovolemic stimuli that increase circulating vasopressin levels also stimulate the release of vasopressin in the lateral septum and lateral ventricle (Demotes-Mainard et al. 1986). Manipulation that facilitates salt intake can also stimulate the systemic release of vasopressin. A prediction from vasopressin's parallel effect is that central administration of vasopressin may actually stimulate salt intake and, indeed, intracerebroventricular (ICV) injections of vasopressin caused a dramatic decrease in NaCl intake in sodium-deficient rats and suppressed sucrose intake. Following that, ICV injection of the V1/V2 receptor antagonist and the V1 receptor antagonist significantly suppressed NaCl intake. But V1 receptor antagonist had no effect on sucrose intake. In contrast, the selective V2 receptor antagonist had no significant effect on NaCl intake. These findings suggest that endogenous vasopressin neurotransmission acting through V1 receptors plays on a role in the amount of salt that is ingested (Flynn et al. 2002).

## 3.2 *Somatostatin*

Somatostatin is a tetradecapeptide originally isolated from the hypothalamus and extensively distributed within both the peripheral and the central nerve system as well as in the pancreas, GI tract, lingual serous glands, and other sites (Johansson et al. 1984; McIntosh et al. 1978; Roberts et al. 1991). It is a classic inhibitory peptide because when it binds to its cognate receptor, it depresses spontaneous discharge rate, hyperpolarizes the postsynaptic membrane, and increases membrane conductance in electrically excitable cells (Bloom 1987; Renaud and Martin 1975). The biological effects of somatostatin are mediated by a family of G-protein-coupled receptors, which are expressed in a tissue-specific manner. Somatostatin receptors have the highest expression levels in jejunum and stomach (Yamada et al. 1993). Somatostatin plays an important role in the regulation of feeding behavior and suppression of good intake (Aponte et al. 1984; Lotter et al. 1981). The reduced feeding activity after somatostatin treatment might depend on the modification of taste. This is supported by reports that some patients treated with somatostatin complain of dysgeusia and lack of appetite (Scalera and Tarozzi 1998). Moreover, somatostatin is found at all levels in taste pathways i.e., the nucleus of the solitary tract, the parabrachial nuclear complex, the parvicellular ventral posterior medialis thalamus, hypothalamus, amygdala, and sensory cortex (Bakhit et al. 1984; De León et al. 1992; Mantyh and Hunt 1984; Moga and Gray 1985) as well as von Ebner's glands of the tongue (Roberts et al. 1991) and saliva (Deville de Periere et al. 1988). Previous work attempted to uncover the function of somatostatin in taste perception. Scalera examined the effects of peripherally administrated somatostatin in protamine zinc to induce a relatively long-lasting effect over the period of fluid intake and taste preference testing in rats. In somatostatin-treated rats, intake of NaCl and sucrose solution decreased, the light/dark cycle of NaCl solution intake was modified, and intake of quinine-HCl and HCl solutions increased significantly (Scalera and Tarozzi 1998). Since somatostatin inhibits the release of growth hormone and the synthesis of protein, taste preference may depend on modification of taste bud morphology and/or size and, indeed, histologic examination showed that taste bud distribution on the tongue appeared to be altered after long-term treatment with somatostatin. Mice injected with somatostatin for 10 days had decreased taste bud density, decreased number of papillae, decreased size, and decreased numbers of taste buds in circumvallate papillae (Scalera and Tarozzi 1998; Scalera 2003). Somatostatin is also known to inhibit the transmission of neural traffic in the gustatory system. Thus, inhibition of the gustatory system might decrease the intake of pleasant solutions (NaCl and sucrose) but enhance that of unpleasant ones (quinine and acid). It seems unlikely that the small changes seen in taste bud morphology and topography are in any way causally or directly related to the effects of somatostatin on taste preferences. Thus, the differences in taste preferences observed after somatostatin treatments might reflect alterations in the central more than in the peripheral taste system. But, it must also be remembered that somatostatin is inhibitory to the release of just about every hormone in the body. The presence of somatostatin receptors on TCs has not yet been described but it is highly likely that somatostatin influences the release of all hormones within TCs and therefore the effect on taste perception as it relates directly to TC reactivity would be the sum of inhibition of all the hormones secreted from TCs.

### 3.3 Oxytocin

Like vasopressin, oxytocin in a nine amino acid peptide that is synthesized in hypothalamic neurons and transported down axons to the posterior pituitary, from where it is ultimately secreted into the bloodstream. It is also secreted within the brain and in a few other tissues, including the ovaries and testes. Oxytocin in the peripheral circulation promotes milk ejection and uterine contractility (Higuchi et al. 1985). CNS-derived oxytocin has been implicated in a variety of behaviors, including the promotion of maternal and social bonding (Insel 1990; Carter et al. 1992), the attenuation of stress and anxiety (Neumann et al. 2000; Windle et al. 1997; Mantella et al. 2003), and inhibition of ingestion (Olson et al. 1991) in rodents. The actions of oxytocin are mediated by specific, high-affinity oxytocin receptors that belong to the rhodopsin-type group of G-protein-coupled receptors, coupling specifically to Gq (Gimpl and Fahrenholz 2001). Oxytocin null mice consume significantly greater daily amounts of 10% sucrose or 0.2% saccharin solutions compared with control littermates when these sweet solutions are freely available (Amico et al. 2005; Billings et al. 2006). Even the mice exposed daily to platform shaker stress consumed more sweetened drinks. A progressive ratio operant licking procedure found that oxytocin null and control mice display a similar motivational drive to consume 10% sucrose solutions and a series of two-bottle intake tests found that oxytocin null mice consume significantly larger volumes of both sweet and non-sweet carbohydrate solutions over observed periods of time (i.e., sucrose, polycose, and cornstarch). The increased sucrose intake of oxytocin null mice seems to result from increased frequency of drinking bouts. The amount consumed during each bout did not differ between the genotypes. These findings indicate that the absence of oxytocin does not affect their appetitive drive to consume sucrose solutions, which would be palatable to the mice. Instead, satiety processes after carbohydrate intake are mediated through neural systems that likely recruit hypothalamic oxytocin neurons and the absence of oxytocin may selectively blunt satiety for carbohydrate-rich foods (Sclafani et al. 2007).

## 4 Hormones That Affect Conditioned Taste Aversion

Conditioned taste aversion (CTA) occurs when association of the taste of a certain food with symptoms caused by a toxic, spoiled, or poisonous substance occurs. Generally, taste aversion is caused after ingestion of the food causes nausea, sickness, or vomiting. The ability to develop a taste aversion is considered an adaptive or survival mechanism that trains the body to avoid or spit out poisonous substances (e.g., poisonous berries) before they can be absorbed and cause bodily harm. This association is meant to prevent the consumption of the same substance (or something that tastes similar) in the future, thus avoiding further poisoning. CTA was discovered when investigators realized that irradiated rats avoided solutions or food that had been present during radiation treatments (Garcia et al. 1955). When rats encountered a novel taste (the conditioning stimulus; CS) and this was followed by transient GI distress caused by low-dose radiation (the unconditioned stimulus; UCS), CTA developed. This response results in a diminished intake of CS upon subsequent presentation. Later studies found that CTA could develop following exposure to a variety of other illness-producing agents, including chemotherapeutic agents, high doses of apomorphine or amphetamine, and lithium chloride (Reilly and Bornovalova 2005). For CTA to develop, the animal must be able to detect the CS; it must be able to become ill from UCS exposure; it must be able to form an association between the UCS and CS; and, finally, it must be able to avoid the CS.

CTA is a relatively simple test to conduct, and it typically requires 2 days of combined training and testing. Mice are placed on food restriction prior to testing. Alternatively, water restriction can be used so that lithium chloride treatment can be paired with the consumption of a saccharin solution. Each animal receives the test substance paired with a flavored saccharin solution, followed by a control treatment (vehicle) paired with the same liquid with a different flavor. After one or more conditioning trials, the animals are then offered a choice between the two flavored liquids. If the animals consume relatively less of the flavored liquid paired with the anorexigenic substance, then it is assumed that the substance produces some degree of malaise.

## 4.1 Peptide $YY_{3\text{-}36}$

Peptide $YY_{3\text{-}36}$ ($PYY_{3\text{-}36}$) is a polypeptide consisting of 36 amino acids and, as described above, is a member of the Y family of peptides. There are two major forms of PYY: $PYY_{1\text{-}36}$ and $PYY_{3\text{-}36}$. However, the most common form of circulating PYY is $PYY_{3\text{-}36}$, which is a ligand for Y2 receptors (Murphy and Bloom 2006). PYY is found in enteroendocrine L cells (the same L cells that synthesize proglucagon from where it is cosecreted with GLP-1 and GLP-2) that are especially abundant in ileum and colon. It is also found in a discrete population of neurons in the brainstem, specifically localized to the gigantocellular reticular nucleus of the medulla oblongata. Like GLP-1 and GLP-2, PYY concentration in the circulation increases after food ingestion and falls during fasting (Murphy et al. 2006). It is reported to slow gastric motility and increase water and electrolyte absorption in the colon (Liu et al. 1996). It may also suppress exocrine pancreatic secretions. Several studies have shown that acute peripheral administration of $PYY_{3\text{-}36}$ inhibits feeding of rodents and primates but Y2 receptor null mice have normal food intake. The anorexia produced by $PYY_{3\text{-}36}$ administration could be due to taste aversion. It is observed that c-Fos immunoreactivity was significantly activated in intermediate nuclei of the solitary tract and area postrema – nuclei known to mediate the response to aversive stimuli – in a dose-dependent manner after peripheral administration of $PYY_{3\text{-}36}$. This shows that $PYY_{3\text{-}36}$ administration produces c-Fos activation in the same brainstem nuclei as lithium chloride. Thus, like lithium chloride, $PYY_{3\text{-}36}$ may reduce food intake by causing a CTA response. In addition, previous experiment shows that 2-h intravenous infusion of $PYY_{3-36}$ using higher doses in both non-food-deprived and food-deprived rats produced a dose-dependent inhibited intake of saccharin solution and a CTA (Chelikani et al. 2006). These results suggest that anorexic doses of $PYY_{3-36}$ may produce a dose-dependent malaise in rats (Batterham et al. 2002; Chelikani et al. 2006), which is similar to that reported for $PYY_{3-36}$ infusion in humans (Batterham et al. 2003; Le Roux et al. 2006). $PYY_{3-36}$ potently inhibits gastric emptying in rats at doses that reduce their food intake (Chelikani et al. 2004), and distention of the gut can produce CTA in rats (Gyetvai and Bárdos 1999; Bárdos 2001). Thus, adverse effects caused by high dose of $PYY_{3-36}$ during eating may be due in part to gastric distention caused by $PYY_{3-36}$-induced slowing of gastric emptying.

## 5 Summary

The study of hormones produced in TCs is now in its infancy. More and more hormones are being uncovered that are produced locally in TCs. Eventually, we should have a taste bud map showing all the hormones, their cellular distribution, and their receptors. While transgenic mice null for certain hormones and their receptors are now available, none are specifically null in just taste buds and uncovering more conclusively the function of the hormones produced locally, than is now possible, will take more time. It is becoming clear, however, that the so-called "gut" hormones involved in food intake and satiety form a chain of cells extending right from colon, small bowel, stomach, taste buds, through to multiple brain areas. Almost certainly, they form a network of signaling molecules that modulates food intake, food perception, quantity and quality of food ingested, feelings of satiety, and metabolic rates.

**Acknowledgements** Our research is funded by the Intramural Program of the NIA/NIH.

## References

Ahrén B, Alumets J, Ericsson M, Fahrenkrug J, Fahrenkrug L, Håkanson R, Hedner P, Lorén I, Melander A, Rerup C, Sundler F (1980) VIP occurs in intrathyroidal nerves and stimulates thyroid hormone secretion. Nature 287:343–345

Amico JA, Vollmer RR, Cai HM, Miedlar JA, Rinaman L (2005) Enhanced initial and sustained intake of sucrose solution in mice with an oxytocin gene deletion. Am J Physiol Regul Integr Comp Physiol 289:R1798–R1806

Aponte G, Leung P, Gross D, Yamada T (1984) Effects of somatostatin on food intake in rats. Life Sci 35:741–746

Bakhit C, Koda L, Benoit R, Morrison JH, Bloom FE (1984) Evidence for selective release of somatostatin-14 and somatostatin-28(1-12) from rat hypothalamus. J Neurosci 4:411–419

Bárdos G (2001) Conditioned taste aversion to gut distension in rats. Physiol Behav 74:407–413

Batterham RL, Cowley MA, Small CJ, Herzog H, Cohen MA, Dakin CL, Wren AM, Brynes AE, Low MJ, Ghatei MA, Cone RD, Bloom SR (2002) Gut hormone PYY(3-36) physiologically inhibits food intake. Nature 418:650–654
Batterham RL, Le Roux CW, Cohen MA, Park AJ, Ellis SM, Patterson M, Frost GS, Ghatei MA, Bloom SR (2003) Pancreatic polypeptide reduces appetite and food intake in humans. J Clin Endocrinol Metab 88:3989–3992
Beinfeld MC, Meyer DK, Eskay RL, Jensen RT, Brownstein MJ (1981) The distribution of cholecystokinin immunoreactivity in the central nervous system of the rat as determined by radioimmunoassay. Brain Res 212:51–57
Bernhardt SJ, Naim M, Zehavi U, Lindemann B (1996) Changes in IP3 and cytosolic Ca2+ in response to sugars and non-sugar sweeteners in transduction of sweet taste in the rat. J Physiol 490:325–336
Billings LB, Spero JA, Vollmer RR, Amico JA (2006) Oxytocin null mice ingest enhanced amounts of sweet solutions during light and dark cycles and during repeated shaker stress. Behav Brain Res 171:134–141
Bjørbaek C, Elmquist JK, Michl P, Ahima RS, van Bueren A, McCall AL, Flier JS (1998) Expression of leptin receptor isoforms in rat brain microvessels. Endocrinology 139:3485–3491
Bloom FE (1987) Molecular diversity and cellular functions of neuropeptides. Prog Brain Res 72:213–220
Branchek T, Smith KE, Walker MW (1998) Molecular biology and pharmacology of galanin receptors. Ann N Y Acad Sci 863:94–107
Branchek TA, Smith KE, Gerald C, Walker MW (2000) Galanin receptor subtypes. Trends Pharmacol Sci 21:109–117
Buffa R, Solcia E, Go VL (1976) Immunohistochemical identification of the cholecystokinin cell in the intestinal mucosa. Gastroenterology 70:528–532
Carter CS, Williams JR, Witt DM, Insel TR (1992) Oxytocin and social bonding. Ann N Y Acad Sci 652:204–211
Chandrashekar J, Hoon MA, Ryba NJ, Zuker CS (2006) The receptors and cells for mammalian taste. Nature 444:288–294
Chase LR, Aurbach GD (1968) Renal adenyl cyclase: anatomically separate sites for parathyroid hormone and vasopressin. Science 159:545–547
Chelikani PK, Haver AC, Reidelberger RD (2004) Comparison of the inhibitory effects of PYY (3-36) and PYY(1-36) on gastric emptying in rats. Am J Physiol Regul Integr Comp Physiol 287:R1064–R1070
Chelikani PK, Haver AC, Reidelberger RD (2006) Dose-dependent effects of peptide YY(3-36) on conditioned taste aversion in rats. Peptides 27:3193–3201
Chen Y, Couvineau A, Laburthe M, Amiranoff B (1992) Solubilization and molecular characterization of active galanin receptors from rat brain. Biochemistry 31:2415–2422
Cooper KE, Kasting NW, Lederis K, Veale WL (1979) Evidence supporting a role for endogenous vasopressin in natural suppression of fever in the sheep. J Physiol 295:33–45
Creutzfeldt W (1979) The incretin concept today. Diabetologia 16:75–85
De León M, Coveñas R, Narváez JA, Tramu G, Aguirre JA, González-Barón S (1992) Distribution of somatostatin-28 (1-12) in the cat brainstem: an immunocytochemical study. Neuropeptides 21:1–11
Demotes-Mainard J, Chauveau J, Rodriguez F, Vincent JD, Poulain DA (1986) Septal release of vasopressin in response to osmotic, hypovolemic and electrical stimulation in rats. Brain Res 381:314–321
Deville de Periere D, Buys-Hillaire D, Favre de Thierrens C, Puech R, Elkaim G, Arancibia S (1988) Somatostatin-immunoreactive concentrations in human saliva and in the submandibular salivary glands of the rat. Possible sexual dependence in the human. J Biol Buccale 16:191–196
Dickson L, Finlayson K (2009) VPAC and PAC receptors: from ligands to function. Pharmacol Ther 121:294–316

Doyle ME, Egan JM (2007) Mechanisms of action of glucagon-like peptide 1 in the pancreas. Pharmacol Ther 113:546–593
Drucker DJ (2005) Biologic actions and therapeutic potential of the proglucagon-derived peptides. Nat Clin Pract Endocrinol Metab 1:22–31
Fahrenkrug J, Schaffalitzky de Muckadell OB, Holst JJ, Jensen SL (1979) Vasoactive intestinal polypeptide in vagally mediated pancreatic secretion of fluid and $HCO_3$. Am J Physiol 237: E535–E540
Farilla L, Hui H, Bertolotto C, Kang E, Bulotta A, Di Mario U, Perfetti R (2002) Glucagon-like peptide-1 promotes islet cell growth and inhibits apoptosis in Zucker diabetic rats. Endocrinology 143:4397–4408
Feng XH, Liu XM, Zhou LH, Wang J, Liu GD (2008) Expression of glucagon-like peptide-1 in the taste buds of rat circumvallate papillae. Acta Histochem 110:151–154
Flynn FW, Kirchner TR, Clinton ME (2002) Brain vasopressin and sodium appetite. Am J Physiol Regul Integr Comp Physiol 282:R1236–R1244
Foucaud M, Archer-Lahlou E, Marco E, Tikhonova IG, Maigret B, Escrieut C, Langer I, Fourmy D (2008) Insights into the binding and activation sites of the receptors for cholecystokinin and gastrin. Regul Pept 145:17–23
Frühbeck G (2006) Intracellular signalling pathways activated by leptin. Biochem J 393:7–20
Garcia J, Kimeldorf DJ, Koelling RA (1955) Conditioned aversion to saccharin resulting from exposure to gamma radiation. Science 122:157–158
Gehlert DR (2004) Introduction to the reviews on neuropeptide Y. Neuropeptides 38:135–140
Gimpl G, Fahrenholz F (2001) The oxytocin receptor system: structure, function, and regulation. Physiol Rev 81:629–683
Goïot H, Attoub S, Kermorgant S, Laigneau JP, Lardeux B, Lehy T, Lewin MJ, Bado A (2001) Antral mucosa expresses functional leptin receptors coupled to STAT-3 signaling, which is involved in the control of gastric secretions in the rat. Gastroenterology 121:1417–1427
Gyetvai B, Bárdos G (1999) Modulation of taste reactivity by intestinal distension in rats. Physiol Behav 66:529–535
Harvey J, Ashford ML (1998) Role of tyrosine phosphorylation in leptin activation of ATP-sensitive K + channels in the rat insulinoma cell line CRI-G1. J Physiol 510:47–61
Herness MS (1989) Vasoactive intestinal peptide-like immunoreactivity in rodent taste cells. Neuroscience 33(2):411–4119
Herness S, Zhao FL (2009) The neuropeptides CCK and NPY and the changing view of cell-to-cell communication in the taste bud. Physiol Behav 97:581–591
Herness S, Zhao FL, Kaya N, Shen T, Lu SG, Cao Y (1995) Distribution of vasoactive intestinal peptide-like immunoreactivity in the taste organs of teleost fish and frog. Histochem J 27:161–165
Herness S, Zhao FL, Lu SG, Kaya N, Shen T (2002) Expression and physiological actions of cholecystokinin in rat taste receptor cells. J Neurosci 22:10018–10029
Herness S, Zhao FL, Kaya N, Shen T, Lu SG, Cao Y (2005) Communication routes within the taste bud by neurotransmitters and neuropeptides. Chem Senses 30(Suppl 1):i37–i38
Higuchi T, Uchide K, Honda K, Negoro H (1985) Functional development of the oxytocin release mechanism and its role in the initiation of parturition in the rat. J Endocrinol 106:311–316
Hoggard N, Mercer JG, Rayner DV, Moar K, Trayhurn P, Williams LM (1997) Localization of leptin receptor mRNA splice variants in murine peripheral tissues by RT-PCR and in situ hybridization. Biochem Biophys Res Commun 232:383–387
Holz GG 4th, Kühtreiber WM, Habener JF (1993) Pancreatic beta-cells are rendered glucose-competent by the insulinotropic hormone glucagon-like peptide-1(7-37). Nature 361:362–365
Iismaa TP, Shine J (1999) Galanin and galanin receptors. Results Probl Cell Differ 26:257–291
Insel TR (1990) Regional changes in brain oxytocin receptors post-partum: time-course and relationship to maternal behaviour. J Neuroendocrinol 2:539–545
Jang HJ, Kokrashvili Z, Theodorakis MJ, Carlson OD, Kim BJ, Zhou J, Kim HH, Xu X, Chan SL, Juhaszova M, Bernier M, Mosinger B, Margolskee RF, Egan JM (2007) Gut-expressed

gustducin and taste receptors regulate secretion of glucagon-like peptide-1. Proc Natl Acad Sci USA 104:15069–15074
Jard S (1988) Mechanisms of action of vasopressin and vasopressin antagonists. Kidney Int Suppl 26:S38–S42
Johansson O, Hökfelt T, Elde RP (1984) Immunohistochemical distribution of somatostatin-like immunoreactivity in the central nervous system of the adult rat. Neuroscience 13:265–339
Kalra SP, Kalra PS (2004) NPY – an endearing journey in search of a neurochemical on/off switch for appetite, sex and reproduction. Peptides 25:465–471
Karatayev O, Baylan J, Leibowitz SF (2009) Increased intake of ethanol and dietary fat in galanin overexpressing mice. Alcohol 43:571–580
Karatayev O, Baylan J, Weed V, Chang S, Wynick D, Leibowitz SF (2010) Galanin knockout mice show disturbances in ethanol consumption and expression of hypothalamic peptides that stimulate ethanol intake. Alcohol Clin Exp Res 34:72–80
Karelson E, Laasik J, Sillard R (1995) Regulation of adenylate cyclase by galanin, neuropeptide Y, secretin and vasoactive intestinal polypeptide in rat frontal cortex, hippocampus and hypothalamus. Neuropeptides 28:21–28
Kawai K, Sugimoto K, Nakashima K, Miura H, Ninomiya Y (2000) Leptin as a modulator of sweet taste sensitivities in mice. Proc Natl Acad Sci USA 97:11044–11049
Koegler FH, Ritter S (1998) Galanin injection into the nucleus of the solitary tract stimulates feeding in rats with lesions of the paraventricular nucleus of the hypothalamus. Physiol Behav 63:521–527
Komatsu R, Matsuyama T, Namba M, Watanabe N, Itoh H, Kono N, Tarui S (1989) Glucagonostatic and insulinotropic action of glucagon like peptide I-(7-36)-amide. Diabetes 38:902–905
Kusakabe T, Matsuda H, Gono Y, Kawakami T, Kurihara K, Tsukuda M, Takenaka T (1998a) Distribution of VIP receptors in the human submandibular gland: an immunohistochemical study. Histol Histopathol 13:373–378
Kusakabe T, Matsuda H, Gono Y, Furukawa M, Hiruma H, Kawakami T, Tsukuda M, Takenaka T (1998b) Immunohistochemical localisation of regulatory neuropeptides in human circumvallate papillae. J Anat 192:557–564
Larsson LI, Rehfeld JF (1979) Localization and molecular heterogeneity of cholecystokinin in the central and peripheral nervous system. Brain Res 165:201–218
le Roux CW, Batterham RL, Aylwin SJ, Patterson M, Borg CM, Wynne KJ, Kent A, Vincent RP, Gardiner J, Ghatei MA, Bloom SR (2006) Attenuated peptide YY release in obese subjects is associated with reduced satiety. Endocrinology 147:3–8
Lee GH, Proenca R, Montez JM, Carroll KM, Darvishzadeh JG, Lee JI, Friedman JM (1996) Abnormal splicing of the leptin receptor in diabetic mice. Nature 379:632–635
Liu CD, Hines OJ, Newton TR, Adrian TE, Zinner MJ, Ashley SW, McFadden DW (1996) Cholecystokinin mediation of colonic absorption via peptide YY: foregut-hindgut axis. World J Surg 20:221–227
Lorén I, Emson PC, Fahrenkrug J, Björklund A, Alumets J, Håkanson R, Sundler F (1979) Distribution of vasoactive intestinal polypeptide in the rat and mouse brain. Neuroscience 4:1953–1976
Lotter EC, Krinsky R, McKay JM, Treneer CM, Porter D Jr, Woods SC (1981) Somatostatin decreases food intake of rats and baboons. J Comp Physiol Psychol 95:278–287
Lu SG, Zhao FL, Herness S (2003) Physiological phenotyping of cholecystokinin-responsive rat taste receptor cells. Neurosci Lett 351:157–160
Lundberg JM, Anggård A, Fahrenkrug J, Hökfelt T, Mutt V (1980) Vasoactive intestinal polypeptide in cholinergic neurons of exocrine glands: functional significance of coexisting transmitters for vasodilation and secretion. Proc Natl Acad Sci USA 77:1651–1655
Mantella RC, Vollmer RR, Li X, Amico JA (2003) Female oxytocin-deficient mice display enhanced anxiety-related behavior. Endocrinology 144:2291–2296
Mantyh PW, Hunt SP (1984) Neuropeptides are present in projection neurons at all levels in visceral and taste pathways: from periphery to sensory cortex. Brain Res 299:297–312

Margetic S, Gazzola C, Pegg GG, Hill RA (2002) Leptin: a review of its peripheral actions and interactions. Int J Obes Relat Metab Disord 26:1407–1433
Martin B, Lopez de Maturana R, Brenneman R, Walent T, Mattson MP, Maudsley S (2005) Class II G-protein-coupled receptors and their ligands in neuronal function and protection. Neuromolecular Med 7:3–36
Martin B, Dotson CD, Shin YK, Ji S, Drucker DJ, Maudsley S, Munger SD (2009) Modulation of taste sensitivity by GLP-1 signaling in taste buds. Ann N Y Acad Sci 1170:98–101
Martin B, Shin YK, White CM, Ji S, Kim W, Carlson OD, Napora JK, Chadwick W, Chapter M, Waschek JA, Mattson MP, Maudsley S, Egan JM (2010) Vasoactive intestinal peptide null mice demonstrate enhanced sweet taste preference, dysglycemia and reduced taste bud leptin receptor expression. Diabetes 59:1143–1152
McCowen KC, Chow JC, Smith RJ (1998) Leptin signaling in the hypothalamus of normal rats in vivo. Endocrinology 139:4442–4447
McDermott BJ, Bell D (2007) NPY and cardiac diseases. Curr Top Med Chem 7:1692–1703
McIntosh C, Arnold R, Bothe E, Becker H, Kobberling J, Creutzfeldt W (1978) Gastrointestinal somatostatin in man and dog. Metabolism 27:1317–1320
Mercer JG, Hoggard N, Williams LM, Lawrence CB, Hannah LT, Morgan PJ, Trayhurn P (1996) Coexpression of leptin receptor and preproneuropeptide Y mRNA in arcuate nucleus of mouse hypothalamus. J Neuroendocrinol 8:733–735
Mitchell V, Bouret S, Prévot V, Jennes L, Beauvillain JC (1999) Evidence for expression of galanin receptor Gal-R1 mRNA in certain gonadotropin releasing hormone neurons of the rostral preoptic area. J Neuroendocrinol 11:805–812
Moga MM, Gray TS (1985) Evidence for corticotropin-releasing factor, neurotensin, and somatostatin in the neural pathway from the central nucleus of the amygdala to the parabrachial nucleus. J Comp Neurol 241:275–284
Moran TH, McHugh PR (1982) Cholecystokinin suppresses food intake by inhibiting gastric emptying. Am J Physiol 242:R491–R497
Murphy KG, Bloom SR (2006) Gut hormones and the regulation of energy homeostasis. Nature 444:854–859
Murphy KG, Dhillo WS, Bloom SR (2006) Gut peptides in the regulation of food intake and energy homeostasis. Endocr Rev 27:719–727
Neumann ID, Krömer SA, Toschi N, Ebner K (2000) Brain oxytocin inhibits the (re)activity of the hypothalamo-pituitary-adrenal axis in male rats: involvement of hypothalamic and limbic brain regions. Regul Pept 96:31–38
Ninomiya Y, Sako N, Imai Y (1995) Enhanced gustatory neural responses to sugars in the diabetic db/db mouse. Am J Physiol 269:R930–R937
Ninomiya Y, Imoto T, Yatabe A, Kawamura S, Nakashima K, Katsukawa H (1998) Enhanced responses of the chorda tympani nerve to nonsugar sweeteners in the diabetic db/db mouse. Am J Physiol 274:R1324–R1330
Olson BR, Drutarosky MD, Chow MS, Hruby VJ, Stricker EM, Verbalis JG (1991) Oxytocin and an oxytocin agonist administered centrally decrease food intake in rats. Peptides 12:113–118
Orskov C, Holst JJ, Knuhtsen S, Baldissera FG, Poulsen SS, Nielsen OV (1986) Glucagon-like peptides GLP-1 and GLP-2, predicted products of the glucagon gene, are secreted separately from pig small intestine but not pancreas. Endocrinology 119:1467–1475
Pedrazzini T, Pralong F, Grouzmann E (2003) Neuropeptide Y: the universal soldier. Cell Mol Life Sci 60:350–377
Perfetti R, Merkel P (2000) Glucagon-like peptide-1: a major regulator of pancreatic beta-cell function. Eur J Endocrinol 143:717–725
Prigeon RL, Quddusi S, Paty B, D'Alessio DA (2003) Suppression of glucose production by GLP-1 independent of islet hormones: a novel extrapancreatic effect. Am J Physiol Endocrinol Metab 285:E701–E707
Rehfeld JF, Friis-Hansen L, Goetze JP, Hansen TV (2007) The biology of cholecystokinin and gastrin peptides. Curr Top Med Chem 7:1154–1165

Reilly S, Bornovalova MA (2005) Conditioned taste aversion and amygdala lesions in the rat: a critical review. Neurosci Biobehav Rev 29:1067–1088
Renaud LP, Martin JB (1975) Electrophysiological studies of connections of hypothalamic ventromedial nucleus neurons in the rat: evidence for a role in neuroendocrine regulation. Brain Res 93:145–151
Roberts IM, Solomon SE, Brusco OA, Goldberg W, Bernstein JJ (1991) Neuromodulators of the lingual von Ebner gland: an immunocytochemical study. Histochemistry 96:153–156
Rosenblum CI, Tota M, Cully D, Smith T, Collum R, Qureshi S, Hess JF, Phillips MS, Hey PJ, Vongs A, Fong TM, Xu L, Chen HY, Smith RG, Schindler C, Van der Ploeg LH (1996) Functional STAT 1 and 3 signaling by the leptin receptor (OB-R); reduced expression of the rat fatty leptin receptor in transfected cells. Endocrinology 137:5178–5181
Said SI, Mutt V (1970) Potent peripheral and splanchnic vasodilator peptide from normal gut. Nature 225:863–864
Sako N, Ninomiya Y, Fukami Y (1996) Analysis of concentration-response relationship for enhanced sugar responses of the chorda tympani nerve in the diabetic db/db mouse. Chem Senses 21:59–63
Scalera G (2003) Peptides that regulate food intake: somatostatin alters intake of amino acid-imbalanced diets and taste buds of tongue in rats. Am J Physiol Regul Integr Comp Physiol 284:R1389–R1398
Scalera G, Tarozzi G (1998) Somatostatin administration alters taste preferences in the rat. Peptides 19:1565–1572
Sclafani A, Rinaman L, Vollmer RR, Amico JA (2007) Oxytocin knockout mice demonstrate enhanced intake of sweet and nonsweet carbohydrate solutions. Am J Physiol Regul Integr Comp Physiol 292:R1828–R1833
Shen T, Kaya N, Zhao FL, Lu SG, Cao Y, Herness S (2005) Co-expression patterns of the neuropeptides vasoactive intestinal peptide and cholecystokinin with the transduction molecules alpha-gustducin and T1R2 in rat taste receptor cells. Neuroscience 130:229–238
Shigemura N, Miura H, Kusakabe Y, Hino A, Ninomiya Y (2003) Expression of leptin receptor (Ob-R) isoforms and signal transducers and activators of transcription (STATs) mRNAs in the mouse taste buds. Arch Histol Cytol 66:253–260
Shigemura N, Ohta R, Kusakabe Y, Miura H, Hino A, Koyano K, Nakashima K, Ninomiya Y (2004) Leptin modulates behavioral responses to sweet substances by influencing peripheral taste structures. Endocrinology 145:839–847
Shin YK, Martin B, Golden E, Dotson CD, Maudsley S, Kim W, Jang HJ, Mattson MP, Drucker DJ, Egan JM, Munger SD (2008) Modulation of taste sensitivity by GLP-1 signaling. J Neurochem 106:455–463
Spanswick D, Smith MA, Groppi VE, Logan SD, Ashford ML (1997) Leptin inhibits hypothalamic neurons by activation of ATP-sensitive potassium channels. Nature 390:521–525
Sperk G, Hamilton T, Colmers WF (2007) Neuropeptide Y in the dentate gyrus. Prog Brain Res 163:285–297
Stoffers DA, Kieffer TJ, Hussain MA, Drucker DJ, Bonner-Weir S, Habener JF, Egan JM (2000) Insulinotropic glucagon-like peptide 1 agonists stimulate expression of homeodomain protein IDX-1 and increase islet size in mouse pancreas. Diabetes 49:741–748
Sun L, Miller RJ (1999) Multiple neuropeptide Y receptors regulate K + and Ca2+ channels in acutely isolated neurons from the rat arcuate nucleus. J Neurophysiol 81:1391–1403
Sun L, Philipson LH, Miller RJ (1998) Regulation of K + and Ca++ channels by a family of neuropeptide Y receptors. J Pharmacol Exp Ther 284:625–632
Takaya K, Ogawa Y, Isse N, Okazaki T, Satoh N, Masuzaki H, Mori K, Tamura N, Hosoda K, Nakao K (1996) Molecular cloning of rat leptin receptor isoform complementary DNAs – identification of a missense mutation in Zucker fatty (fa/fa) rats. Biochem Biophys Res Commun 225:75–83
Tatemoto K, Rökaeus A, Jörnvall H, McDonald TJ, Mutt V (1983) Galanin – a novel biologically active peptide from porcine intestine. FEBS Lett 164:124–128

Vaisse C, Halaas JL, Horvath CM, Darnell JE Jr, Stoffel M, Friedman JM (1996) Leptin activation of Stat3 in the hypothalamus of wild-type and ob/ob mice but not db/db mice. Nat Genet 14:95–97

Wang S, Ghibaudi L, Hashemi T, He C, Strader C, Bayne M, Davis H, Hwa JJ (1998a) The GalR2 galanin receptor mediates galanin-induced jejunal contraction, but not feeding behavior, in the rat: differentiation of central and peripheral effects of receptor subtype activation. FEBS Lett 434:277–282

Wang S, Hashemi T, Fried S, Clemmons AL, Hawes BE (1998b) Differential intracellular signaling of the GalR1 and GalR2 galanin receptor subtypes. Biochemistry 37:6711–6717

Wang X, Cahill CM, Piñeyro MA, Zhou J, Doyle ME, Egan JM (1999) Glucagon-like peptide-1 regulates the beta cell transcription factor, PDX-1, in insulinoma cells. Endocrinology 140:4904–4907

Waters SM, Krause JE (2000) Distribution of galanin-1, -2 and -3 receptor messenger RNAs in central and peripheral rat tissues. Neuroscience 95:265–271

Wheway J, Herzog H, Mackay F (2007) NPY and receptors in immune and inflammatory diseases. Curr Top Med Chem 7:1743–1752

Willms B, Werner J, Holst JJ, Orskov C, Creutzfeldt W, Nauck MA (1996) Gastric emptying, glucose responses, and insulin secretion after a liquid test meal: effects of exogenous glucagon-like peptide-1 (GLP-1)-(7-36) amide in type 2 (noninsulin-dependent) diabetic patients. J Clin Endocrinol Metab 81:327–332

Windle RJ, Shanks N, Lightman SL, Ingram CD (1997) Central oxytocin administration reduces stress-induced corticosterone release and anxiety behavior in rats. Endocrinology 138: 2829–2834

Witt M (1995) Distribution of vasoactive intestinal peptide-like immunoreactivity in the taste organs of teleost fish and frog. Histochem J 27(2):161–165

Wraith A, Törnsten A, Chardon P, Harbitz I, Chowdhary BP, Andersson L, Lundin LG, Larhammar D (2000) Evolution of the neuropeptide Y receptor family: gene and chromosome duplications deduced from the cloning and mapping of the five receptor subtype genes in pig. Genome Res 10(3):302–310

Xu ZQ, Shi TJ, Hökfelt T (1996) Expression of galanin and a galanin receptor in several sensory systems and bone anlage of rat embryos. Proc Natl Acad Sci USA 93:14901–14905

Yamada Y, Kagimoto S, Kubota A, Yasuda K, Masuda K, Someya Y, Ihara Y, Li Q, Imura H, Seino S (1993) Cloning, functional expression and pharmacological characterization of a fourth (hSSTR4) and a fifth (hSSTR5) human somatostatin receptor subtype. Biochem Biophys Res Commun 195:844–852

Yamashita T, Murakami T, Otani S, Kuwajima M, Shima K (1998) Leptin receptor signal transduction: OBRa and OBRb of fa type. Biochem Biophys Res Commun 246:752–759

Zander M, Madsbad S, Madsen JL, Holst JJ (2002) Effect of 6-week course of glucagon-like peptide 1 on glycaemic control, insulin sensitivity, and beta-cell function in type 2 diabetes: a parallel-group study. Lancet 359:824–830

Zhao FL, Shen T, Kaya N, Lu SG, Cao Y, Herness S (2005) Expression, physiological action, and coexpression patterns of neuropeptide Y in rat taste-bud cells. Proc Natl Acad Sci USA 102:11100–11105

# Endocannabinoid Modulation in the Olfactory Epithelium

**Esther Breunig, Dirk Czesnik, Fabiana Piscitelli, Vincenzo Di Marzo, Ivan Manzini, and Detlev Schild**

**Abstract** Appetite, food intake, and energy balance are closely linked to the endocannabinoid system in the central nervous system. Now, endocannabinoid modulation has been discovered in the peripheral olfactory system of larval *Xenopus laevis*. The endocannabinoid 2-AG has been shown to influence odorant-detection thresholds according to the hunger state of the animal. Hungry animals have increased 2-AG levels due to enhanced synthesis of 2-AG in sustentacular supporting cells. This renders olfactory receptor neurons, exhibiting CB1 receptors, more

E. Breunig (✉)
Department of Neurophysiology and Cellular Biophysics, University of Göttingen, 37073 Göttingen, Germany
DFG Cluster of Excellence 171, Humboldtallee 23, 37073 Göttingen, Germany
e-mail: ebreuni@gwdg.de

D. Czesnik
Department of Neurophysiology and Cellular Biophysics, University of Göttingen, 37073 Göttingen, Germany
e-mail: dczesni@gwdg.de

F. Piscitelli and V. Di Marzo
Endocannabinoid Research Group, Institute of Biomolecular Chemistry, Consiglio Nazionale delle Ricerche, Via Campi Flegrei 34, Pozzuoli (NA), Italy
e-mail: fpiscitelli@icb.cnr.it, vdimazo@icmib.na.cnr.it

I. Manzini
Department of Neurophysiology and Cellular Biophysics, University of Göttingen, 37073 Göttingen, Germany
DFG Research Center for Molecular Physiology of the Brain (CMPB), Humboldtallee 23, 37073 Göttingen, Germany
e-mail: imanzin@gwdg.de

D. Schild
Department of Neurophysiology and Cellular Biophysics, University of Göttingen, 37073 Göttingen, Germany
DFG Cluster of Excellence 171, Humboldtallee 23, 37073 Göttingen, Germany
DFG Research Center for Molecular Physiology of the Brain (CMPB), Humboldtallee 23, 37073 Göttingen, Germany
e-mail: dschild@gwdg.de

W. Meyerhof et al. (eds.), *Sensory and Metabolic Control of Energy Balance*, Results and Problems in Cell Differentiation 52,
DOI 10.1007/978-3-642-14426-4_11, 

sensitive at detecting lower odorant concentrations, which probably helps the animal to locate food. Since taste and vision are also influenced by endocannabinoids, this kind of modulation might boost sensory inputs of food in hungry animals.

## 1 Introduction

Appetite stimulation is a well-known effect of cannabis abuse. The reason for this is that active compounds of the cannabis plant bind to endogenous receptors in the brain. The most abundant cannabinoid receptor in the central nervous system is known as the CB1 receptor (Matsuda et al. 1990). Using current techniques, CB1 receptors could be localized in brain regions including those responsible for the control of energy balance, e.g., in nuclei of the mesolimbic pathways and the hypothalamus (Herkenham et al. 1991; Tsou et al. 1998). Furthermore, the CB1 receptor is expressed in peripheral tissues involved in energy homeostasis like adipose tissue, liver, gastrointestinal tract, pancreas, thyroid gland, and adrenal gland (Pagotto et al. 2006; Demuth and Mollemann 2006; Juan-Picó et al. 2006).

Shortly after identifying the endogenous cannabinoid receptor and its distribution, it has been shown that the endocannabinoid system has the ability of controlling appetite, food intake, and energy balance (Matias and Di Marzo 2007; Osei-Hyiaman et al. 2006; Horvath 2006). Furthermore, endogenous ligands of cannabinoid receptors, known as endocannabinoids, modulate metabolic functions in cells of peripheral tissues, such as adipocytes, hepatocytes, and the gastrointestinal tract (Pagotto et al. 2006). Consistently, selective inverse agonists or antagonists of CB1 receptors reduce weight and can be used for the treatment of obesity (Kirkham and Tucci 2006; Di Marzo and Després 2009).

Another key mechanism for the regulation of appetite is the sensory properties of food. Taste, smell, aspect, and texture of food might be rewarding and thus facilitate their choice and consumption. For food seeking, the sense of smell is particularly important. A variety of animals use odors to locate and identify food. Furthermore, olfactory sensitivity is increased when animals are hungry (Duclaux et al. 1973; Rolls 2005; Aimé et al. 2007). The mechanisms underlying the sensory control of food intake are currently under investigation. Since endocannabinoids regulate many central and peripheral processes related to energy balance, they seem likely to link also sensory systems, particularly olfaction, to food intake.

## 2 Endocannabinoid System in Peripheral Sensory Systems

### *2.1 Dorsal Root Ganglion Cells, Photoreceptor Cells, and Taste Cells*

Several sensory systems have been shown to be modulated by the endocannabinoid system in the periphery. For example, in the retina, cannabinoids speed up the dynamics of the phototransduction deactivation cascade in the outer segment of cones (Straiker et al. 1999; Struik et al. 2006). Furthermore, CB1 receptors are expressed on type II taste cells, coexpressing the T1R3 sweet taste receptor component. Endocannabinoids act here as enhancers of sweet taste (Yoshida et al. 2010). In addition, CB1 receptors are expressed on dorsal root ganglion cells (Hohmann and Herkenham 1999; Bridges et al. 2003) and may play a role in the spinal nociceptive system (Morisset et al. 2001; Agarwal et al. 2007).

## 2.2 *Olfactory Receptor Neurons*

Although since 1991, CB1 receptors have been known to be located in the olfactory bulb (Herkenham et al. 1991; Cesa et al. 2001; Egertová and Elphick 2000), first hints for the presence of the endocannabinoid system in the peripheral olfactory system – the olfactory epithelium – were found only recently. A study of Migliarini et al. (2006) demonstrated the presence of CB1 receptor mRNA in the olfactory epithelium of *Xenopus laevis* tadpoles.

In 2007, we could localize CB1 receptors on dendrites of olfactory receptor neurons (ORNs) of the same species (Fig. 1a and b) and revealed that ORNs are modulated by cannabinoids (Czesnik et al. 2007). In order to describe this modulatory effect, we examined $[Ca^{2+}]_i$ transients of ORNs evoked by odorants. These odorant-induced $[Ca^{2+}]_i$ transients turned out to be an excellent tool for this purpose, since the highly reproducible increase in $[Ca^{2+}]_i$ in somata of ORNs is a measure for the activity of these cells (Fig. 1c, black trace). CB1 receptor antagonists like AM251, AM281, and LY320135 reduced the amplitude and delayed the onset of odorant-induced responses (Fig. 1c, red trace). A subsequent treatment of the tissue with the CB1 receptor agonist HU210 accelerated the recovery of these responses (Fig. 1c, green trace).

This study strongly indicated the presence of an endocannabinoid system in the olfactory epithelium, which was investigated in detail in a subsequent study (Breunig et al. 2010). 2-AG could be identified as an endocannabinoid acting on ORNs. Blockers of the diacylglycerol lipases (RHC80267, orlistat, OMDM-187, OMDM-188), which are the main synthesizing enzymes of 2-AG (Bisogno et al. 2003), had the same effect on odorant-induced $[Ca^{2+}]_i$ transients as CB1 receptor antagonists in the study by Czesnik et al. (2007). They reduced and delayed odorant-induced responses of ORNs. As in our previous study, the CB1 receptor agonist HU210 could restore the response behavior quickly.

The sources of 2-AG reside in both sustentacular supporting cells and ORNs. 2-AG is synthesized in supporting cells by diacylglycerol lipase α and in ORNs by diacylglycerol lipase β. Thus, 2-AG acts on CB1 receptors on ORN dendrites in a paracrine as well as an autocrine manner. The synthesis in sustentacular supporting cells but not in ORNs was found to depend on hunger. In line with this, food deprivation of animals significantly increased diacylglycerol lipase α mRNA expression in sustentacular supporting cells (Fig. 1d), leading to about 1.5-fold increase of 2-AG concentration in the olfactory epithelium (Fig. 1e). The finding of

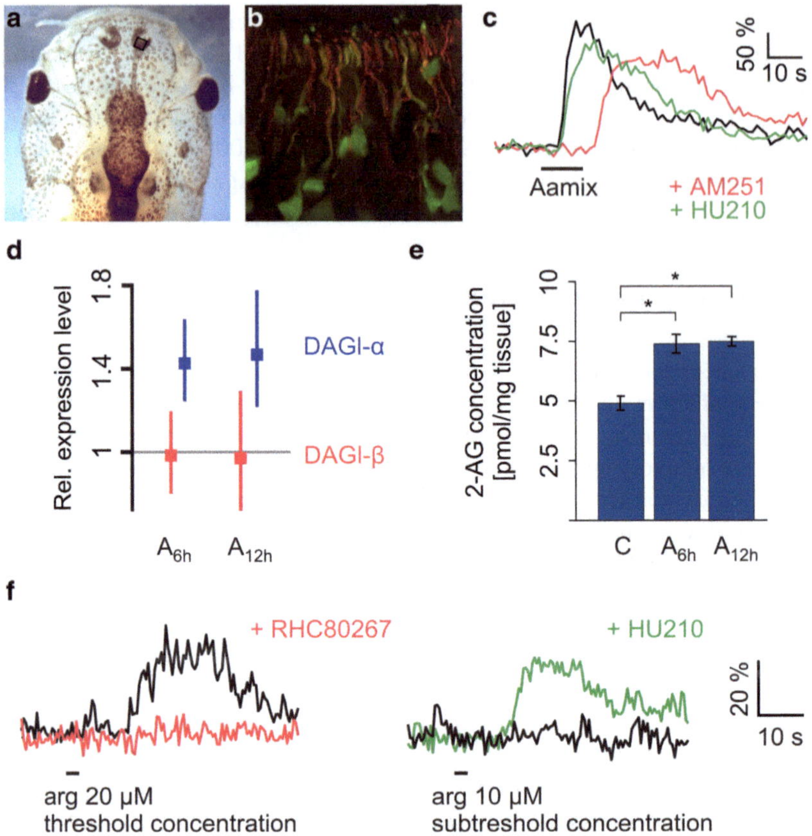


**Fig. 1** Endocannabinoid action in the olfactory epithelium. (**a**) Larval *Xenopus laevis*. (**b**) Higher magnification of the olfactory epithelium, marked in (**a**), ORNs (*green*, biocytin–avidin–stain) double stained with anti-CB1 antibodies (*red*). (**c**) Aamix-evoked $[Ca^{2+}]_i$ transients in somata of individual ORNs (*black trace*) were reduced and delayed after wash-in of the CB1 receptor antagonist AM251 (5 μM, *red trace*). Two minutes after HU210 wash-in (10 μM), the $[Ca^{2+}]_i$ transients recovered almost completely (*green traces*). The *black line* below the response traces indicates the application of the odorants. (**d**) The relative expression levels of DAG lipase α in sustentacular cells (*blue points*), but not DAG lipase β in ORNs (*red points*), were enhanced after food deprivation ($A_{6h}$, food deprived for 6 h, $A_{12h}$, food deprived for 12 h). The *bars* depict the S.E.M. (**e**) The concentration of 2-AG increased to about 1.5-fold upon food deprivation. $*p < 0.05$. (**f**) The *black traces* present ORN $[Ca^{2+}]_i$ responses to threshold (20 μM) and subthreshold concentrations (10 μM) of arg. The response at the detection threshold was abolished after addition of RHC80267 (50 μM, *red trace*) to the bath solution. Vice versa, subthreshold odorant concentrations led to an odorant response after addition of HU210 (10 μM, 2 min, *green trace*) to the bath. Figure adapted from Czesnik et al. (2007) and Breunig et al. (2010)

the hunger-induced elevation of endocannabinoid levels in organs deputed to mediate food intake, both at the sensory and at the higher levels, is not unprecedented in mammals and nonmammalian vertebrates and has been observed so far in the hypothalamus and limbic forebrain of food-deprived rats (Kirkham et al. 2002), as well as in the telencephalon of the songbird zebra finch (Soderstrom et al. 2004) and two fish species: the goldfish *Carassius auratus* (Valenti et al. 2005) and in the brain of sea bream *Sparus aurata* (Piccinetti et al. 2010).

To assess the physiological benefit of this system in the olfactory epithelium, reduced levels of 2-AG in the olfactory epithelium were obtained by 2-AG synthesis blockage with the diacylglycerol lipase inhibitors RHC80267 or orlistat, and the sensitivity of ORNs was measured under this condition. Under a suppression of 2-AG synthesis, higher odorant concentrations were required to elicit responses in individual ORNs (Fig. 1f, left). Vice versa, CB1 receptor activation with the specific CB1 receptor agonist HU210, which mimics higher 2-AG levels, decreased odorant threshold concentrations (Fig. 1f, right).

These data show that the feeding state of larval *Xenopus laevis* alters the sensitivity of individual ORNs by the orexigenic modulator 2-AG. Hungry larvae

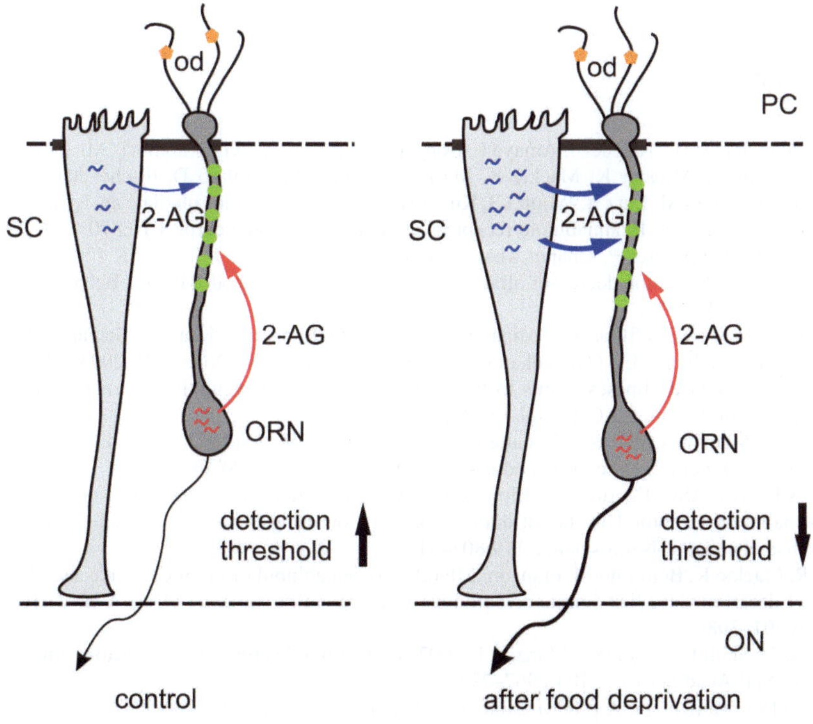


**Fig. 2** Overview of the role of the endocannabinoid system in the olfactory epithelium. Under control conditions, tonic levels of 2-AG are synthesized. DAG lipase α mRNA synthesis in sustentacular cells (~, *blue*) is enhanced upon food deprivation and leads to enhanced levels of 2-AG, activating CB1 receptors (*green spots*) on ORN dendrites (*blue arrows*) in a paracrine manner. This renders ORNs more sensitive. DAG lipase β mRNA expression in ORNs (~, *red*) is not affected by food deprivation. 2-AG synthesized in ORNs feeds back on ORNs (*red arrow*) in an autocrine manner. *PC* principal cavity; *ON* olfactory nerve; *od* odorant. Figure adapted from Breunig et al. (2010)

have higher 2-AG levels in their olfactory epithelium. Thus, lower concentrations of odorants can be perceived, most probably making it easier for tadpoles to locate food (Fig. 2).

## 3 Conclusion

The sensory modalities taste, smell, aspect, and texture influence food intake. Already at the most peripheral stage of the olfactory system, the endocannabinoid 2-AG acts as an orexigenic mediator and renders ORNs more sensitive. Since endocannabinoids also enhance sweet taste and modulate vision, it might be that all peripheral sensory systems are tuned by endocannabinoids and that hunger enhances not only smell but also taste, aspect, and texture of food.

## References


Agarwal N, Pacher P, Tegeder I, Amaya F, Constantin CE, Brenner GJ, Rubino T, Michalski CW, Marsicano G, Monory K, Mackie K, Marian C, Batkai S, Parolaro D, Fischer MJ, Reeh P, Kunos G, Kress M, Lutz B, Woolf CJ, Kuner R (2007) Cannabinoids mediate analgesia largely via peripheral type 1 cannabinoid receptors in nociceptors. Nat Neurosci 10:870–879

Aimé P, Duchamp-Viret P, Chaput MA, Savigner A, Mahfouz M, Julliard AK (2007) Fasting increases and satiation decreases olfactory detection for a neutral odor in rats. Behav Brain Res 179:258–264

Bisogno T, Howell F, Williams G, Minassi A, Cascio MG, Ligresti A, Matias I, Schiano-Moriello A, Paul P, Williams EJ, Gangadharan U, Hobbs C, Di Marzo V, Doherty P (2003) Cloning of the first sn1-DAG lipases points to the spatial and temporal regulation of endocannabinoid signaling in the brain. J Cell Biol 163:463–468

Breunig E, Manzini I, Piscitelli F, Gutermann B, Di Marzo V, Schild D, Czesnik D (2010) The endocannabinoid 2-AG controls odor sensitivity. J Neurosci 30:8965–8973

Bridges D, Rice ASC, Egertová M, Elphick MR, Winter J, Michael GJ (2003) Localization of the cannabinoid receptor 1 in the rat dorsal root ganglion using in situ hybridization and immunohistochemistry. Neuroscience 119:803–812

Cesa R, Mackie K, Beltramo M, Franzoni MF (2001) Cannabinoid receptor CB1-like and glutamic acid decarboxylase-like immunoreactivities in the brain of *Xenopus laevis*. Cell Tissue Res 306:391–398

Czesnik D, Schild D, Kuduz J, Manzini I (2007) Cannabinoid action in the olfactory epithelium. Proc Natl Acad Sci USA 104:2967–2972

Demuth DG, Mollemann A (2006) Cannabinoid signaling. Life Sci 78:549–563

Di Marzo V, Després JP (2009) CB1 antagonists for obesity – what lessons have we learned from rimonabant? Nat Rev Endocrinol 5:633–638

Duclaux R, Feisthauer J, Cabanac M (1973) Effects of a meal on the pleasantness of food and nonfood odors in man. Physiol Behav 10:1029–1033

Egertová M, Elphick MR (2000) Localisation of cannabinoid receptors in the rat brain using antibodies to the intracellular C-terminal tail of CB. J Comp Neurol 422:159–171

Herkenham M, Lynn AB, Johnson MR, Melvin LS, de Costa BR, Rice KC (1991) Characterization and localization of cannabinoid receptors in rat brain: a quantitative in vitro autoradiographic study. J Neurosci 11:563–583

Hohmann AG, Herkenham M (1999) Localization of central cannabinoid CB1 receptor messenger RNA in neuronal subpopulations of rat dorsal root ganglia: a double-label in situ hybridization study. Neuroscience 90:923–931

Horvath TL (2006) The unfolding cannabinoid story on energy homeostasis: central or peripheral site of action? Int J Obes (Lond) 30:30–32

Juan-Picó P, Fuentes E, Bermúdez-Silva FJ, Javier Díaz-Molina F, Ripoll C, Rodríguez de Fonseca F, Nadal A (2006) Cannabinoid receptors regulate Ca(2+) signals and insulin secretion in pancreatic beta-cell. Cell Calcium 39:155–162

Kirkham TC, Tucci SA (2006) Endocannabinoids in appetite control and the treatment of obesity. CNS Neurol Disord Drug Targets 5:272–292

Kirkham TC, Williams CM, Fezza F, Di Marzo V (2002) Endocannabinoid levels in rat limbic forebrain and hypothalamus in relation to fasting, feeding and satiation: stimulation of eating by 2-arachidonoyl glycerol. Br J Pharmacol 136:550–557

Matias I, Di Marzo V (2007) Endocannabinoids and the control of energy balance. Trends Endocrinol Metab 18:27–37

Matsuda LA, Lolait SJ, Brownstein MJ, Young AC, Bonner TI (1990) Structure of a cannabinoid receptor and functional expression of the cloned cDNA. Nature 346:561–564

Migliarini B, Marucci G, Ghelfi F, Carnevali O (2006) Endocannabinoid system in *Xenopus laevis* development: CB1 receptor dynamics. FEBS Lett 580:1941–1945

Morisset V, Ahluwalia J, Nagy I, Urban L (2001) Possible mechanisms of cannabinoid-induced antinociception in the spinal cord. Eur J Pharmacol 429:93–100

Osei-Hyiaman D, Harvey-White J, Bátkai S, Kunos G (2006) The role of the endocannabinoid system in the control of energy homeostasis. Int J Obes (Lond) 30:33–38

Pagotto U, Marsicano G, Cota D, Lutz B, Pasquali R (2006) The emerging role of the endocannabinoid system in endocrine regulation and energy balance. Endocr Rev 27:73–100

Piccinetti CC, Migliarini B, Petrosino S, Di Marzo V, Carnevali O (2010) Anandamide and AM251 via water, modulate food intake at central and peripheral level in fish. Gen Comp Endocrinol 166:259–267

Rolls ET (2005) Taste, olfactory, and food texture processing in the brain, and the control of food intake. Physiol Behav 85:45–56

Soderstrom K, Tian Q, Valenti M, Di Marzo V (2004) Endocannabinoids link feeding state and auditory perception-related gene expression. J Neurosci 24:10013–10021

Straiker A, Stella N, Piomelli D, Mackie K, Karten HJ, Maguire G (1999) Cannabinoid CB1 receptors and ligands in vertebrate retina: localization and function of an endogenous signaling system. Proc Natl Acad Sci U S A 96:14565–14570

Struik ML, Yazulla S, Kamermans M (2006) Cannabinoid agonist WIN 55212-2 speeds up the cone response to light offset in goldfish retina. Vis Neurosci 23:285–293

Tsou K, Brown S, Sañudo-Peña MC, Mackie K, Walter JM (1998) Immunohistochemical distribution of cannabinoid receptors in the rat central nervous system. Neuroscience 83:393–411

Valenti M, Cottone E, Martinez R, De Pedro N, Rubio M, Viveros MP, Franzon MF, Delgado MJ, Di Marzo V (2005) The endocannabinoid system in the brain of *Carassius auratus* and its possible role in the control of food intake. J Neurochem 95:662–672

Yoshida R, Ohkuri T, Jyotaki M, Yasuo T, Horio N, Yasumatsu K, Sanematsu K, Shigemura N, Yamamoto T, Margolskee RF, Ninomiya Y (2010) Endocannabinoids selectively enhance sweet taste. Proc Natl Acad Sci U S A 107:935–939

# The Olfactory Bulb: A Metabolic Sensor of Brain Insulin and Glucose Concentrations via a Voltage-Gated Potassium Channel

**Kristal Tucker, Melissa Ann Cavallin, Patrick Jean-Baptiste, K.C. Biju, James Michael Overton, Paola Pedarzani, and Debra Ann Fadool**

**Abstract** The voltage-gated potassium channel, Kv1.3, contributes a large proportion of the current in mitral cell neurons of the olfactory bulb where it assists to time the firing patterns of action potentials as spike clusters that are important for odorant detection. Gene-targeted deletion of the Kv1.3 channel, produces a "super-smeller" phenotype, whereby mice are additionally resistant to diet- and genetically-induced obesity. As assessed via an electrophysiological slice preparation of the olfactory bulb, Kv1.3 is modulated via energetically important molecules – such as insulin and glucose – contributing to the body's metabolic response to fat intake. We discuss a biophysical characterization of modulated synaptic communication in the slice following acute glucose and insulin stimulation, chronic elevation of insulin in mice that are in a conscious state, and induction of diet-induced obesity. We have discovered that Kv1.3 contributes an unusual nonconducting role – the detection of metabolic state.

K. Tucker, M.A. Cavallin, P. Jean-Baptiste, K. Biju and J.M. Overton
Program in Neuroscience, The Florida State University, Tallahassee, FL, USA

P. Pedarzani
Research Department of Neuroscience, Physiology and Pharmacology, University of College London, London, UK

D.A. Fadool (✉)
Program in Neuroscience, The Florida State University, Tallahassee, FL, USA
Institute of Molecular Biophysics, The Florida State University, Tallahassee, FL, USA
e-mail: dfadool@bio.fsu.edu

W. Meyerhof et al. (eds.), *Sensory and Metabolic Control of Energy Balance*,
Results and Problems in Cell Differentiation 52,
DOI 10.1007/978-3-642-14426-4_12, 

# 1 Introduction

## *1.1 Kv1.3 Channel Distribution and Function*

The voltage-dependent potassium channel, Kv1.3, is a mammalian homolog of the *Shaker* subfamily of potassium channels, which has a selective distribution within the nervous system including high expression in the dentate gyrus, the olfactory bulb, and the olfactory cortex (Kues and Wunder 1992). The biophysical properties of the channel were first described as characterized in T lymphocytes (Cahalan et al. 1985), where today, active drug discovery efforts to find the most effective molecules to block the vestibule of the channel remain a focus of intensive research designed to dampen inflammatory responses associated with degenerative diseases, principally multiple sclerosis (Cahalan and Chandy 2009). Although, classically, one envisions potassium channels as dampeners of excitability through timing of the interspike interval (ISI) and shape of the action potential, as well as drivers for setting the resting membrane potential (Jan and Jan 1994; Yellen 2002), recent data have demonstrated that this particular potassium channel has a plethora of nonconductive functions that make it highly unusual, or at least untraditional (Kaczmarek 2006). One of the reasons that Kv1.3 may have multiple regulatory roles could be attributed to its structure and favorability as a central scaffold upon which signaling molecules build protein–protein interactions. Kv1.3 has 17 tyrosine residues, several of which lie within good recognition motifs for tyrosine phosphorylation (Pawson 1995; Huganir and Jahn 2000). Site-directed mutagenesis has been applied to both the channel and predicted regulatory kinases and adaptor proteins to map signaling cascades, associated with modulating channel function (Holmes et al. 1996a, b; Bowlby et al. 1997; Fadool et al. 1997; Fadool and Levitan 1998; Cook and Fadool 2002; Colley et al. 2004, 2007, 2009; Marks and Fadool 2007). For example, the cellular tyrosine kinase, src, phosphorylates residues $Tyr^{137}$ and $Tyr^{449}$ and has been found to substantially suppress Kv1.3 current, while slowing the kinetics of inactivation (Cook and Fadool 2002), while the receptor-linked epidermal growth factor receptor phosphorylates only $Tyr^{479}$ and predominantly acts to speed the kinetics of inactivation with only minor reduction in current amplitude (Bowlby et al. 1997). In the olfactory bulb, Kv1.3 is a substrate for phosphorylation by the insulin receptor kinase, whereby stimulation with the ligand insulin evokes no change in kinetic properties of the channel, but a reduction in current magnitude attributed to a reduction in mean open probability and not unitary conduction (Fadool and Levitan 1998; Fadool et al. 2000).

The discovery of the many nonconductive roles for Kv1.3 was made through loss of function studies using a whole-animal, targeted deletion of the Kv1.3 gene (Koni et al. 2003). Other laboratories including us noticed that the Kv1.3-null mice were thinner than their wild-type counterparts without caloric self-restriction (Fig. 1a and b) (Xu et al. 2003, 2004; Fadool et al. 2004). Using a custom designed metabolic chamber to quantify systems physiology parameters and ingestive behaviors (Fig. 3a–b) (Williams et al. 2003), we found that the Kv1.3-null animals, more

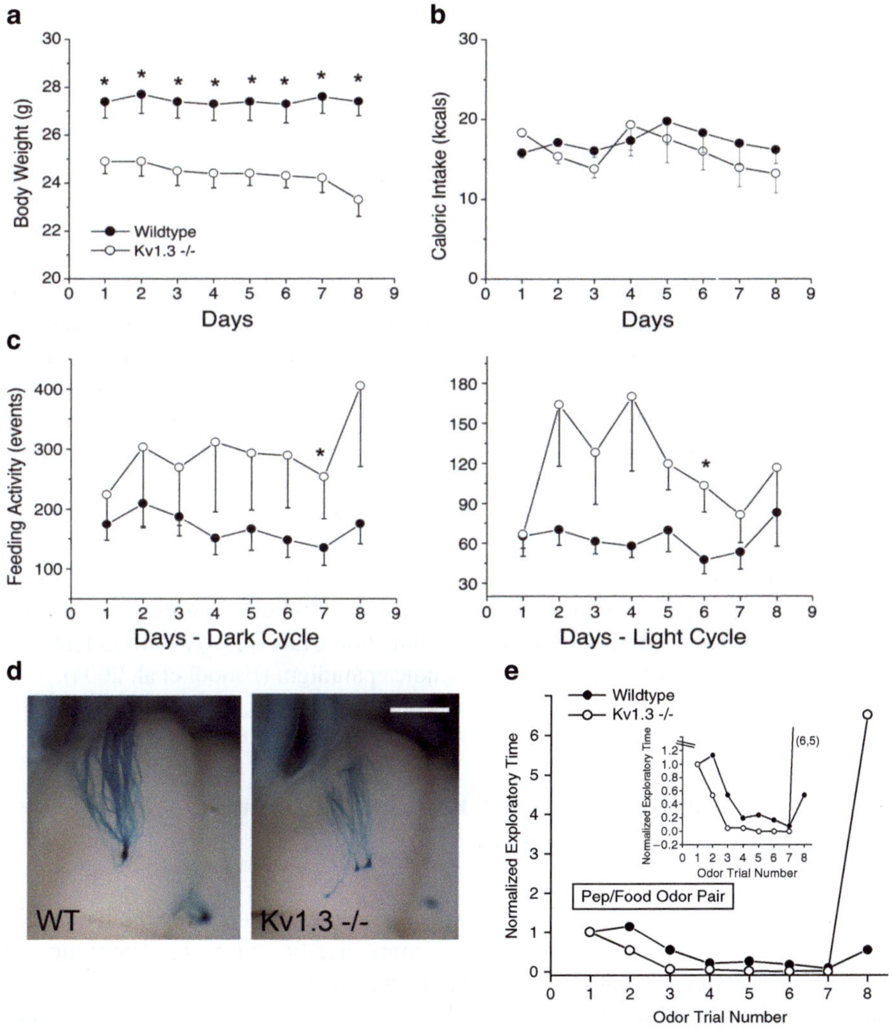


**Fig. 1** Loss of Kv1.3 gene causes a reduction in body weight, modified ingestive behaviors, disruption in axonal targeting in the olfactory bulb, and increased olfactory discrimination in mice. (**a**) Line graph of the mean ± standard error of the mean (s.e.m.) body weight monitored for ten mice of each genotype. Wildtype = control C57Bl6 mice, Kv1.3-/- = mice with gene-targeted deletion of the Kv1.3 ion channel. (**b**) Line graph of the mean ± s.e.m. caloric intake for ten mice of each genotype monitored for 8 days. (**c**) Line graph of the mean ± s.e.m. feeding activity for ten mice of each genotype monitored for 8 days during the 12 h dark cycle (*left*) or 10 h of the light cycle (*right*). Computerized monitoring was disrupted for a 2 h interval/day for cage maintenance. *Asterisk* = significantly different by Student's *t*-test at the 95% confidence level. (**a–c**) Reproduced with permission from Fadool et al. (2004). (**d**) Axonal projections are visualized in a whole-mount of the olfactory bulb in M72*ires*tauLacZ mice maintained on a wildtype (WT) or Kv1.3-null (Kv1.3-/-) background. Note the supernumerary glomerular projection in the Kv1.3-null animal at P20 that will remain unpruned through late adult (>2 years) (Biju et al. 2008). Scale bar = 1 mm. (**e**) Mice with a gene-targeted deletion (Kv1.3-/-) have an increased olfactory discrimination based upon enhanced performance in an odor-habituation paradigm. *Inset* = expanded *Y* axis to better visualize habituation phase (**a–c**, **e**). Reproduced with permission from Fadool et al. (2004)

frequently broke a photobeam that guarded access to their food receptacles (Fig. 1c), and oppositely, less frequently attended the water on a lick-o-meter (data not shown), while still maintaining identical total calorie and water intake as that of wild-type animals. The null animals had a slightly elevated metabolic activity and an increased locomotor activity particularly in the dark cycle (Fadool et al. 2004). Interestingly, Hennige et al. (2009) has demonstrated that the i.c.v. injection of the Kv1.3 pore blocker, margatoxin, similarly elevates locomotor activity and increases cortical action potential frequency.

Since Kv1.3 carries 60–80% of the outward current in the olfactory bulb primary output neurons (Fadool and Levitan 1998; Colley et al. 2004), the mitral cells, we were intrigued to explore olfactory-related phenotypes in the gene-targeted deleted models. By breeding the Kv1.3-null mice, onto a background of mice with a genetic marker for particular classes of odorant receptor-identified olfactory sensory neurons, we were able to discern that the projections of neurons into the olfactory bulb no longer converged to a single glomerular synaptic unit, but rather were supernumerary in target (Fig. 1d) (Biju et al. 2008). Within a given glomerulus, subsequent dual-color fluorescent confocal microscopy studies demonstrated that glomeruli were no longer homogenous, but rather contained sensory projections from more than one class of olfactory sensory neurons (Biju et al. 2008). Behaviorally, the Kv1.3-null mice, had an increased olfactory ability in terms of both discrimination of molecular features of odorants, determined by odor-habituation trials (Fig. 1e) and in terms of odorant threshold, determined by the two-choice paradigm (Fadool et al. 2004).

## 2 Mechanistic Link Between Kv1.3 Ion Channel, Metabolism, and Olfaction

Given the world-wide health epidemic of the rise of the incidence of type II diabetes and unwanted weight gain (obesity), we immediately sought to determine the relationship between metabolic disorders, energy homeostasis, the modulation of this channel by insulin, and olfaction. We decided to challenge the Kv1.3-null animals, with a moderately high fat (MHF; 32% fat) diet for a period of 26 weeks and quantify body weight gain, serum chemistry, and metabolic profile as previously described by Tucker et al. (2008). Unlike wild-type counterparts, Kv1.3-null animals did not deposit significant quantities of fat in typical locations and were resistant to increases in body weight over the test interval (Fig. 2d). Wild-type animals demonstrated the induction of prediabetic blood chemistry (Fig. 2a), unlike that of Kv1.3-null animals, in which basal and fat challenged fasting glucose, serum insulin, and serum leptin levels were significantly reduced (data not shown). Using intranasal insulin delivery across the cribiform plate, into the olfactory bulb, we demonstrated that animals maintained on a MHF-diet now failed to exhibit an increase in insulin-induced Kv1.3 phosphorylation, developing a degree of insulin resistance at the level of the ion channel (Fig. 2b) (Marks et al. 2009).

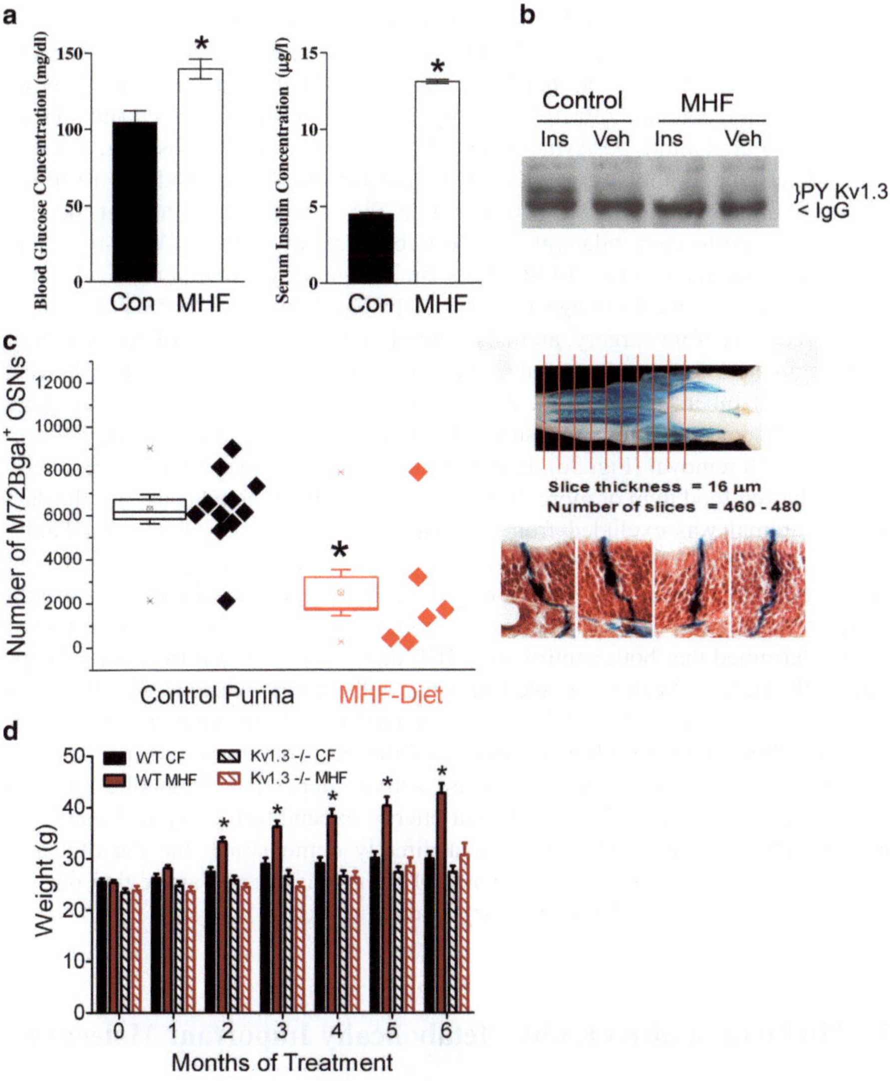


**Fig. 2** Mice maintained on a moderately high-fat (MHF) diet develop a prediabetic blood chemistry, resistance to Kv1.3 channel phosphorylation, and a loss of an OR-identified class of olfactory sensory neurons. Mice with a gene-targeted loss of Kv1.3 ion channel are resistant to obesity. (**a**) *Bar graph* of blood glucose and serum insulin concentrations for six wild-type mice maintained for 52 weeks on a control Purina chow (Con) or 32% fat diet (MHF). (**b**) Same cohort of mice in which mice were intranasally administered saline vehicle (Veh) or 0.1 µg/ml insulin twice daily for 8 days. Proteins were immunoprecipitated with an antibody directed against Kv1.3 protein, separated by SDS-PAGE, and then probed with an antibody that recognizes tyrosine specific phosphorylation (PY Kv1.3). IgG = immunoglobulin band. (**c**) Combined scatter (each mouse) and box plot (population mean and s.e.m.) of the number of M72 B-galactosidase positive neurons in the epithelia of mice maintained on different dietary regimes. Same experimental diet paradigm was performed (as in **a** and **b**) on mice with a genetic marker for the M72 odorant

When genetically-identified odor receptor tagged mice were placed on a MHF-diet, and then a number of OR-specified olfactory sensory neurons were counted across the whole epithelia, we found that there was a loss of half of the neurons, or more directly, half the potential olfactory sensory information being received and relayed to the Kv1.3-containing postsynaptic targets, the mitral cell neurons (Fig. 2c).

To determine if the loss of Kv1.3, in the olfactory bulb and the resulting enhanced olfactory ability were responsible for the resistance to diet-induced obesity, we performed bilateral olfactory bulbectomy (OBX). Wild-type and Kv1.3-null animals underwent OBX (or sham) surgery by bilateral removal of the olfactory bulbs at 9 weeks of age as described by Getchell et al. (2005). Following a 2 week recovery from surgery, animals were placed on either control Purina diet or MHF regime for 5–6 weeks and then monitored for 8 days in the custom-housed metabolic chambers (Fig. 3a–b). At the end of the 16 week study, mice were behaviorally confirmed to be anosmic and then sacrificed to anatomically confirm complete bulb removal (Fig. 3c). If an OBX-treated animal was found to be able to detect a buried food item or more than 25% of the bulb remained (Fig. 3d), the data for that animal was excluded from the data set for analysis. Quite remarkably, OBX-treated, Kv1.3-null animals were no longer able to abrogate weight gain following maintenance on the MHF-diet (Fig. 3f). Figure 3e demonstrates weight gain in OBX-treated wild-type animals in comparison (Fig. 3e). Metabolic assessment determined that both control and MHF-diet fed Kv1.3-null treatment groups transiently increased caloric intake following bulbectomy, whereas wild-type animals, did not. In particular, MHF-diet challenged Kv1.3-null mice increased their basal metabolic rate. Combined removal of the olfactory bulb and maintenance on the MHF-diet, was found to decrease activity-dependent metabolic rate and thereby decrease total weight-dependent energy expenditure computed using the Weir equation (Weir 1949). These data directly demonstrate that the olfactory bulb contributes to the metabolic balance of energy usage; a brain region outside of the traditional hypothalamic pituitary, endocrine axis.

## 3 Modulation of Kv1.3 by Metabolically Important Molecules

If gene-targeted deletion of Kv1.3 channel evokes a thin, supersmeller phenotype that is resistant to diet- and genetically-induced obesity, and maintenance of wild-type mice on high fat diets with presumably elevated glucose and insulin levels

**Fig. 2** (continued) receptor, M72*ires*tauLacZ. Each whole epithelia were sectioned in entirety and then processed for β-galactosidase product to identify M72 expressing olfactory sensory neurons (OSNs). *Neutral red* was utilized as a counterstain (*right*) to better resolve OSNs in context. (**d**) *Bar graph* of the mean body weight ± s.e.m. of wild-type (WT) or Kv1.3-null (Kv1.3-/-) mice maintained for 26 weeks on either the control Purina chow (CF) or 32% fat diet (MHF) (**a** and **b**). Reproduced with permission from Marks et al. (2009). (**c**) Whole-mount photograph modified with permission from Biju et al. (2008)

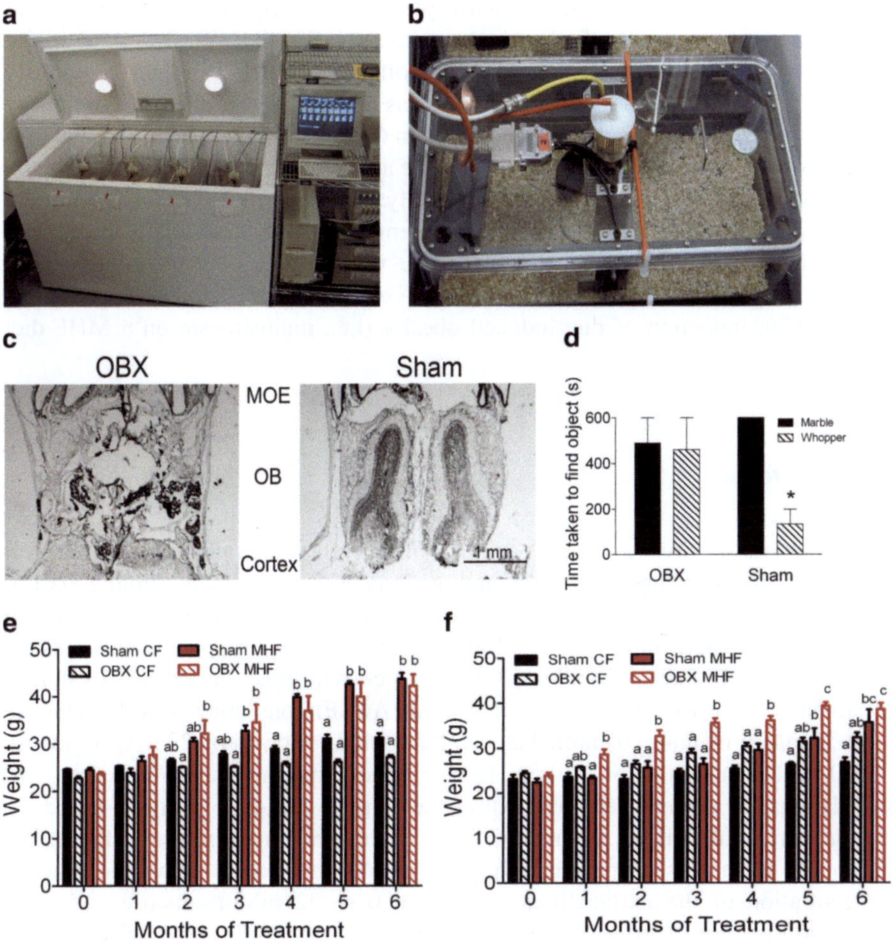


**Fig. 3** Removal of the olfactory bulb in Kv1.3-null mice restores their sensitivity to diet-induced obesity via a reduction in energy expenditure. (**a**) Photograph showing the custom engineered metabolic chamber that is automated to collect respiratory quotient, locomotor activity, ingestive behavior every 30 s for 8 days while regulating circadian rhythms. (**b**) Close up photograph of the cage insert of the metabolic chamber that demonstrates how the cage is aerated, temperature regulated, and sealed to acquire indirect measures of calorimetry. (**c**) Photomicrograph of a 16 μM thick coronal cyrosection through the olfactory bulb which was histologically stained to confirm complete surgical oblation of the olfactory bulb. (**d**) *Bar graph* of the mean ± s.e.m. retrieval time for mice to uncover a scented object. OBX = mice with olfactory bulbectomy, SHAM = mice undergoing cranial surgery but bulb intact. (**e**) *Bar graph* of the mean ± s.e.m. body weight for wild-type mice undergoing OBX or SHAM surgery and placed on a Purina control chow (CF) or 32% fat diet (MHF) for 14 weeks. (**f**) Same as panel e but for Kv1.3-null mice.
Note: mice that were not visually confirmed as successfully ablated (panel **c**) or behaviorally anosmic (panel **d**), were not included in the weight study (panels **e–f**)

decreases the number of olfactory sensory neurons, then, what is the functional ramification at the level of electrical excitability for the mitral cell; a major contributor of Kv1.3 conductance in the olfactory bulb? We had previously reported biochemical evidence that Kv1.3 was a substrate for phosphorylation by insulin using a heterologous expression system (Fadool and Levitan 1998), and thus used this same system to determine if glucose also could modulate Kv1.3 biophysics. In order to test whether two metabolically important molecules – insulin and glucose – modulated Kv1.3 in vivo, it was essential for us to additionally develop an adult olfactory bulb slice preparation so that we could explore modulation after chronic stimulation with these molecules (i.e., intranasal delivery approaches) or following induction of diet-induced obesity (i.e., maintenance on a MHF-diet since birth).

## *3.1 Glucose*

Acute glucose sensitivity of olfactory bulb mitral cells was evaluated by whole-cell current-clamp recordings from horizontal sections (325 μm) prepared from C57BL/6 mice (wildtype) or mice with a Kv1.3 gene-targeted deletion (Kv1.3-null). Mitral cell membrane potentials were held at −65 mV to prevent spontaneous spiking followed by a 4 s, perithreshold (50–100 pA) current injection every 20 s during treatment with artificial cerebral spinal fluid (ACSF) containing 0 mM D-glucose with 22 mM D-mannitol osmotic balance for 10 min followed by 22 mM D-glucose and 0 mM D-mannitol for 10 min. During these experiments, we observed two populations of glucose sensitive mitral cells from wild-type animals based on changes in total spiking frequency. Forty-eight percent of mitral cells tested, exhibited an increased spiking frequency in response to changing the glucose concentration of the extracellular bath from 0 to 22 mM D-glucose and were therefore considered to be glucose excited. The other 52% exhibited a drop in spiking frequency, or were glucose inhibited, in the presence of 22 mM D-glucose. Mitral cells from Kv1.3-null mice, however, exhibited no change in spiking frequency due to change in glucose concentration. This suggests that Kv1.3 expression is important, at least in part, for glucose sensitivity of mitral cells.

## *3.2 Insulin*

Acute application of insulin to mitral cells shortens the ISI as determined through the Gaussian fitting of ISI histograms generated from action potentials evoked from current injections stepped from 25 to 500 pA in cells held near the resting membrane potential. Mitral cell firing frequency linearly increased from 10 to approximately 45 Hz over current steps ranging from 25 to 200 pA. Following acute insulin stimulation of the slice for 20 min, the firing frequency significantly increased from

25 to 60 Hz in response to the same current steps. Interestingly, at stronger current injections, ranging from 300 to 500 pA, firing frequency in untreated mitral cells progressively fell below 45 Hz due to spike adaptation, but following acute insulin stimulation, mitral cells could maintain firing rates up to 85 Hz without adaptation. Spike shape was significantly modified following acute insulin stimulation, whereby the action potential width was reduced, the action potential amplitude was increased, and the spike decay time ($1/e$) was faster. At perithreshold current injections (5–30 pA) using long duration current steps (5,000 ms) we found that the characteristic spike clustering generated by mitral cells was modified following acute insulin stimulation. Spike clustering is due to intrinsic membrane properties, persists in the presence of NBQX and APV synaptic blockers, and is thought to provide frequency information for odorant discrimination (Balu and Strowbridge 2007). We found that the pause duration of the spike clusters was significantly decreased following insulin stimulation. If insulin were delivered chronically as opposed to acutely, then a different pattern of spike clustering was observed. We intranasally delivered insulin, twice a day for 8 days, as per Marks et al. (2009), to P50 and older animals, and then measured generated action potentials evoked at perithreshold current injections. Following chronic insulin treatment, mitral cells exhibited two basal types of firing frequencies that were discreetly opposite in graphed activity patterns using raster plots. Basally, neurons either had extremely high levels of spike clusters with short pause durations, or neurons fired with short latency to first action potential spike and only a single spike cluster of short duration was observed. Following application of insulin to these slices, insulin now evoked a decrease in the action potential firing frequency, regardless of which initial pattern of activity was exemplified. Finally, mice that were placed on a MHF-diet via feeding the dam prior to pairing the parents, and then retaining weaned pups on the diet through adulthood (P35–P65), showed basal mitral cell properties that included modified timing of spike clusters, spike train adaption, and partial firing. Acute application of insulin to animals maintained on the MHF-diet, since birth was now ineffective in changing action potential firing frequencies.

## 4 Conclusion of Nonconductive Roles for Kv1.3 Governing Energy Homeostasis

We have demonstrated that disruption of the Kv1.3 gene, results in reduced body weight, abrogation of obesity, modified axonal targeting in the olfactory system, increased olfactory ability, and changes in serum blood chemistry. Maintenance on a moderately high-fat diet reduces the number of olfactory sensory neurons while elevating insulin and glucose that we have directly shown to alter mitral cell biophysical properties in a slice configuration of the olfactory bulb. A variant in the promoter of the Kv1.3 gene (i.e., gain in channel function), and referred to as the diabetes risk allele, has recently been associated with impaired glucose tolerance, lower insulin sensitivity, higher fasting plasma glucose, and impaired olfactory dysfunction in males (Tschritter et al. 2006; Guthoff et al. 2009). It appears that natural changes in the sensitivity of the OB driven by modulation of Kv1.3 (in rats and humans) may contribute to the body's metabolic response to fat intake or energy imbalance.

**Acknowledgments** We would like to thank Mr. Michael Henderson and Steven J. Godbey for routine technical assistance and mouse colony husbandry. We would like to thank Ms. Marita Madson for many insightful electrophysiological discussions. We would like to thank Mr. Charles Badland for artistic assistance in the visuals used in our oral presentation for this symposium. This work was supported by NIH grants R01 DC003387 & F31 DC010097 from the NIDCD, the Tallahassee Memorial Hospital/Robinson Foundation, and a Sabbatical Award from Florida State University.

## References

Balu R, Strowbridge BW (2007) Opposing inward and outward conductances regulate rebound discharges in olfactory mitral cells. J Neurophysiol 97:1959–1968

Biju KC, Marks DR, Mast TG, Fadool DA (2008) Deletion of voltage-gated channel affects glomerular refinement and odorant receptor expression in the mouse olfactory system. J Comp Neurol 506:161–179

Bowlby MR, Fadool DA, Holmes TC, Levitan IB (1997) Modulation of the Kv1.3 potassium channel by receptor tyrosine kinases. J Gen Physiol 110:601–610

Cahalan MD, Chandy KG (2009) The functional network of ion channels in T lymphocytes. Immunol Rev 231:59–87

Cahalan MD, Chandy KG, DeCoursey TE, Gupta S (1985) A voltage-gated potassium channel in human T lymphocytes. J Physiol 358:197–237

Colley B, Biju KC, Visegrady A, Campbell S, Fadool DA (2007) TrkB increases Kv1.3 ion channel half-life and surface expression. Neuroscience 144:531–546

Colley B, Cavallin MA, Biju KC, Fadool DA (2009) Brain-derived neurotrophic factor modulation of Kv1.3 channel is dysregulated by adaptor proteins Grb10 and nShc. Neuroscience 144(2): 531–46

Colley B, Tucker K, Fadool DA (2004) Comparison of modulation of Kv1.3 channel by two receptor tyrosine kinases in olfactory bulb neurons of rodents. Receptors Channels 10:25–36

Cook KK, Fadool DA (2002) Two adaptor proteins differentially modulate the phosphorylation and biophysics of Kv1.3 ion channel by SRC kinase. J Biol Chem 277:13268–13280

Fadool DA, Holmes TC, Berman K, Dagan D, Levitan IB (1997) Multiple effects of tyrosine phosphorylation on a voltage-dependent potassium channel. J Neurophysiol 78:1563–1573

Fadool DA, Levitan IB (1998) Modulation of olfactory bulb neuron potassium current by tyrosine phosphorylation. J Neurosci 18:6126–6137

Fadool DA, Tucker K, Perkins R, Fasciani G, Thompson RN, Parsons AD, Overton JM, Koni PA, Flavell RA, Kaczmarek LK (2004) Kv1.3 channel gene-targeted deletion produces "Super-Smeller Mice" with altered glomeruli, interacting scaffolding proteins, and biophysics. Neuron 41:389–404

Fadool DA, Tucker K, Phillips JJ, Simmen JA (2000) Brain insulin receptor causes activity-dependent current suppression in the olfactory bulb through multiple phosphorylation of Kv1.3. J Neurophysiol 83:2332–2348

Getchell TV, Liu H, Vaishnav RA, Kwong K, Stromberg AJ, Getchell ML (2005) Temporal profiling of gene expression during neurogenesis and remodeling in the olfactory epithelium at short intervals after target ablation. J Neurosci Res 80(3):309–329

Guthoff M, Tschritter O, Berg D, Liepelt I, Schulte C, Machicao F, Haering HU, Fritsche A (2009) Effect of genetic variation in Kv1.3 on olfactory function. Diabetes Metab Res Rev 25:523–527

Hennige AM, Sartorius T, Lutz SZ, Tschritter O, Preissl H, Hopp S, Fritsche A, Rammensee HG, Ruth P, Häring H-U (2009) Insulin-mediated cortical activity in the slow frequency range is diminished in obese mice and promotes physical inactivity. Diabetologia 52:2416–2424

Holmes TC, Fadool DA, Levitan IB (1996a) Tyrosine phosphorylation of the Kv1.3 potassium channel. J Neurosci 16:1581–1590

Holmes TC, Fadool DA, Ren R, Levitan IB (1996b) Association of src tyrosine kinase with a human potassium channel mediated by SH3 domain. Science 274:2089–2091

Huganir RL, Jahn R (2000) Signalling mechanisms. Curr Opin Neurobiol 10:289–292

Jan LY, Jan NJ (1994) Potassium channels and their evolving gates. Nature (London) 371:119–122

Kaczmarek LK (2006) Non-conducting functions of voltage-gated ion channels. Nat Rev Neurosci 7:761–771

Koni PA, Khanna R, Chang MC, Tang MD, Kaczmarek LK, Schlichter LC, Flavella RA (2003) Compensatory anion currents in Kv1.3 channel-deficient thymocytes. J Biol Chem 278: 39443–39451

Kues WA, Wunder F (1992) Heterogeneous expression patterns of mammalian potassium channel genes in developing and adult rat brain. Eur J Neurosci 4:1296–1308

Marks DR, Fadool DA (2007) Post-synaptic density 95 (PSD-95) affects insulin-induced Kv1.3 channel modulation of the olfactory bulb. J Neurochem 103:1608–1627

Marks DR, Tucker K, Cavallin MA, Mast TG, Fadool DA (2009) Awake intranasal insulin delivery modifies protein complexes and alters memory, anxiety, and olfactory behaviors. J Neurosci 29:6734–6751

Pawson T (1995) Protein modules and signalling networks. Nature (London) 373(573-580):1995

Tucker K, Overton JM, Fadool DA (2008) Kv1.3 gene-targeted deletion alters longevity and reduces adiposity by increasing locomotion and metabolism in melanocortin-4 receptor-null mice. Int J Obes 32:1222–1232

Tschritter O, Machicao F, Stefan N, Schäfer S, Weigert C, Staiger H, Spieth C, Häring H-U, Fritsche A (2006) A new variant in the human Kv1.3 gene is associated with low insulin sensitivity and impaired glucose tolerance. J Clin Endocr Metab 91:654–658

Weir JB (1949) New methods for calculating metabolic rate with special reference to protein metabolism. J Physiol 109:1–9

Williams TD, Chambers JB, Gagnon SP, Roberts LM, Henderson RP, Overton JM (2003) Cardiovascular and metabolic responses to fasting and thermoneutrality in Ay mice. Physiol Behav 78:615–623

Xu J, Koni PA, Wang P, Li G, Kaczmarek L, Wu Y, Li Y, Flavell RA, Desir GV (2003) The voltage-gated potassium channel Kv1.3 regulates energy homeostasis and body weight. Hum Mol Genet 12(5):551–559

Xu J, Wang P, Li Y, Li G, Kaczmarek LK, Wu Y, Koni PA, Flavell RA, Desir GV (2004) The voltage-gated potassium channel Kv1.3 regulates peripheral insulin sensitivity. Proc Natl Acad Sci USA 101:3112–3117

Yellen G (2002) The voltage-gated potassium channels and their relatives. Nature (London) 419:35–42

# Energy Homeostasis Regulation in *Drosophila*: A Lipocentric Perspective

**Ronald P. Kühnlein**

**Abstract** The fruit fly *Drosophila* is a centenarian in research service, but a novice as an invertebrate model system for energy homeostasis research. The last couple of years, however, witnessed numerous technical advances driving the rise of this model organism in central areas of energy balance research such as food perception, feeding control, energy flux and lipometabolism. These studies demonstrate an unanticipated evolutionary conservation of genes and mechanisms governing central aspects of energy homeostasis. Accordingly, research on *Drosophila* promises both, a systems biology view on the regulatory network, which governs lifelong energy control in a complex eukaryotic organism as well as, important insights into the mammalian energy balance control with a potential impact on the diagnostic and therapeutic strategies in the treatment of human lipopathologies such as obesity.

## 1 Introduction: The *Drosophila* Model Organism in Energy Homeostasis Research

In spite of its small size of only few millimeters a century of research career made the fruit fly *Drosophila melanogaster* a giant among the animal model organisms. Since its introduction into the laboratory by Noble laureate Thomas H. Morgan in the early twentieth century, *Drosophila* proved of value as a versatile, cost-effective, robust and particularly fast genetic model system. While in the early days the fly helped elucidating the basic mechanisms of inheritance, the following decades witnessed important contributions of *Drosophila* to fundamental questions in various other biological disciplines such as developmental and cell biology,

R.P. Kühnlein
Max-Planck-Institut für biophysikalische Chemie, Forschungsgruppe Molekulare Physiologie, Am Fassberg 11, 37077 Göttingen, Germany
e-mail: rkuehnl@gwdg.de

W. Meyerhof et al. (eds.), *Sensory and Metabolic Control of Energy Balance*, Results and Problems in Cell Differentiation 52,
DOI 10.1007/978-3-642-14426-4_13, 

neurobiology and behaviour and lately as invertebrate model organism for metabolic homeostasis research (Baker and Thummel 2007; Leopold and Perrimon 2007).

Key to the lasting success of this model system is the comprehensive and ever-growing methodological repertoire, which also granted the flies leap from single gene analysis to the "-omics" era. The driving force behind both scientific discovery and technological advances in *Drosophila* is a globally active fly research community whose proceedings are brought together in the excellent *Drosophila* research database called FlyBase (Tweedie et al. 2009; www.flybase.org).

Candidate gene identification in complex regulatory phenomena such as energy homeostasis requires a systems biology approach (Fig. 1), combined with a quantifiable lead phenotype e.g. storage fat content. Public stock centres offer thousands of mutant and transgenic fly strains for forward and reverse genetic screens, which represent the core of functional analysis (St Johnston 2002). Genome-wide RNAi technology in *Drosophila* cell lines (Beller et al. 2008; Guo et al. 2008) and in vivo (Pospisilik et al. 2010; Kühnlein unpublished) identified a number of fat storage regulators in the fly. Importantly, the fly system allows to address a potential energy balance function even for essential genes by combining gene knockdown techniques with switchable in vivo expression systems (McGuire et al. 2004) or by recombination-based genetic mutant mosaic analysis (Wu and Luo 2006).

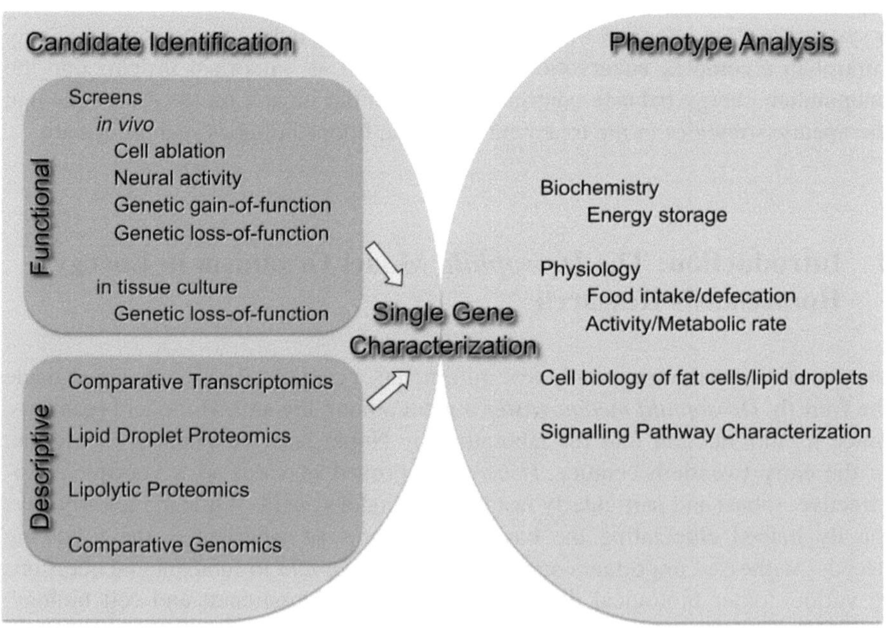


**Fig. 1** Strategies of energy homeostasis regulator identification in *Drosophila*. Candidate gene identification is based on numerous functional approaches in vivo and in tissue culture as well as on descriptive large-scale techniques. Functional single gene characterization by phenotyping a broad range of biochemical, physiological, cell biological and genetic parameters confirm energy balance regulators in vivo (for details see text)

The cellular basis of energy homeostasis control has been addressed by in vivo cell ablation of insulin producing cells in the central brain (Rulifson et al. 2002) or of the neuroendocrine corpora cardiaca cells of the ring gland (Lee and Park 2004). Recently, the neuronal wiring of fat storage control (Al-Anzi et al. 2009) and taste-induced behaviour (Gordon and Scott 2009) in the fly has been analyzed by modulation of neural activity in distinct neuronal subpopulations of the central nervous system.

Although functional approaches are the first choice to identify energy homeostasis regulators, more descriptive methods have been successfully applied as well. For example, comparative transcriptomics of fed and starved flies at different ontogenetic stages has been applied to identify central regulators of fat storage control such as *sugarless* (Zinke et al. 2002) and the Brummer lipase (Grönke et al. 2005), the orthologue of mammalian adipose triglyceride lipase (ATGL) (Haemmerle et al. 2006; Zimmermann et al. 2004). Lipid droplet proteomics (Beller et al. 2006; Cermelli et al. 2006) and lipolytic proteomics (Birner-Grünberger and Kühnlein unpublished) are two more strategies, which make use of intracellular localization and enzymatic activity, respectively, to identify energy balance regulator candidates. Moreover, comparative genomics in silico aims at the identification of fly orthologs of mammalian energy homeostasis genes and at the description of evolutionarily conserved metabolic pathways. Needless to say, that all descriptive candidate identification strategies require subsequent functional single gene analysis to promote a candidate gene to a *bona fide* energy homeostasis regulator.

*Drosophila* body fat content is the lead phenotype used to identify energy homeostasis regulators, as it is accessible not only to a single gene but also to high-throughput analysis. To this aim, enzymatic (Grönke et al. 2003; Pospisilik et al. 2010 and Kühnlein unpublished) or chromatography-based (Al-Anzi et al. 2009) determination of fly homogenate fat content has been used. Moreover lipid droplet size and number monitoring by microscopic inspection (Guo et al. 2008) or by an image segmentation-based optical readout (Beller et al. 2008) in tissue culture has been applied. Recently developed noninvasive imaging methods, such as MRI (Null et al. 2008; Righi et al. 2010), mesoscopic fluorescence tomography (Vinegoni et al. 2008) or multispectral optoacoustic tomography (Ma et al. 2009) have been downscaled to the *Drosophila* dimensions and promise lipid composition and fat depot-specific storage analysis in intact flies.

Next to lipids carbohydrates represent like in mammals, the second column of organismal energy storage in flies. Enzymatic methods are currently in use to determine tissue glycogen storage as well as circulating sugar concentrations (Grönke et al. 2007; Lee and Park 2004). However, mass spectroscopy-based metabolome analysis in the fly (Kamleh et al. 2009) will not only allow identification and quantification of complex carbohydrates but also lipometabolism signatures.

Comprehensive energy flux analysis in *Drosophila* has long been hampered by the unavailability of suitable methods to quantitatively trace energy intake and expenditure in the fly. However, recent advances in fly food intake quantification (Ja et al. 2007; Wong et al. 2009) allowed the description of a *Drosophila* circadian feeding rhythm (Xu et al. 2008). Integration of food intake quantification with defecation analysis and activity/metabolic rate measurements (Van Voorhies et al. 2008; www.trikinetics.com), combined with fly food of defined composition (Bass et al. 2007; Grandison et al. 2009) now promises seamless monitoring of energy traffic in this model organism.

In summary, technological advances improved phenotypic analysis of fly mutants to an extent, which makes the powerful genetics of *Drosophila* e.g. epistasis analysis for signalling pathway identification in vivo, fully applicable to energy homeostasis research (Fig. 1).

## 2 Fly Organs Governing Organismal Energy Homeostasis

Doubtlessly organismal energy homeostasis control is a regulatory achievement of all cells in the animal. Yet certain cells, tissues or organs play a prominent role in the energy management from food perception and intake to macronutrient partitioning, fuel storage and remobilization, to energy consumption and waste excretion. The intimate link between different organs in energy homeostasis control is exemplified by the *takeout (to)* gene, which is expressed in gustatory neurons of the taste sensilla and in the fat body, the major energy storage organ of the fly (see below). Mutants, lacking *to* suffer from hyperphagia and fat body hypertrophy under *ad libitum* feeding conditions, show sensory neuron desensitizing and impaired foraging behaviour in response to starvation (Meunier et al. 2007). Also the importance of circadian feeding rhythms and behavioural control has been demonstrated as peripheral clock inhibition in gustatory neurons causes hyperactivity, hyperphagia and increased energy storage in flies (Chatterjee et al. 2010). The importance of neural gene expression for energy balance control, is exemplified by the cGMP-dependent protein kinase Foraging, which increases sugar-responsiveness in flies (Belay et al. 2007) or by the cAMP responsive transcription factor CREB, which controls energy storage (Iijima et al. 2009). Recent advances, in the understanding of the anatomical network underlying feeding behaviour in the fly have been reviewed in Buch and Pankratz (2009) and Melcher et al. (2007) and emphasize the importance of the suboesophageal ganglion (SOG) of the central nervous system as nodal relay for feeding coordination (Fig. 2).

Once the activity of gustatory receptors in chemosensory sensilla located predominantly in the mouthparts and the legs of the adult fly have triggered feeding behaviour, the meal enters the alimentary canal and gets temporarily stored in the crop (Fig. 2). Little is known about the digestive properties of the fly crop and the regulatory processes, which govern the further transport of the food to the midgut, which is the major site of digestion and nutrient absorption. A recent study identified the midgut-expressed orphan nuclear receptor DHR96 as important regulator of dietary fat metabolism, likely by transcriptional control of a gastric lipase (Sieber and Thummel 2009). Lipid droplet accumulation in defined regions

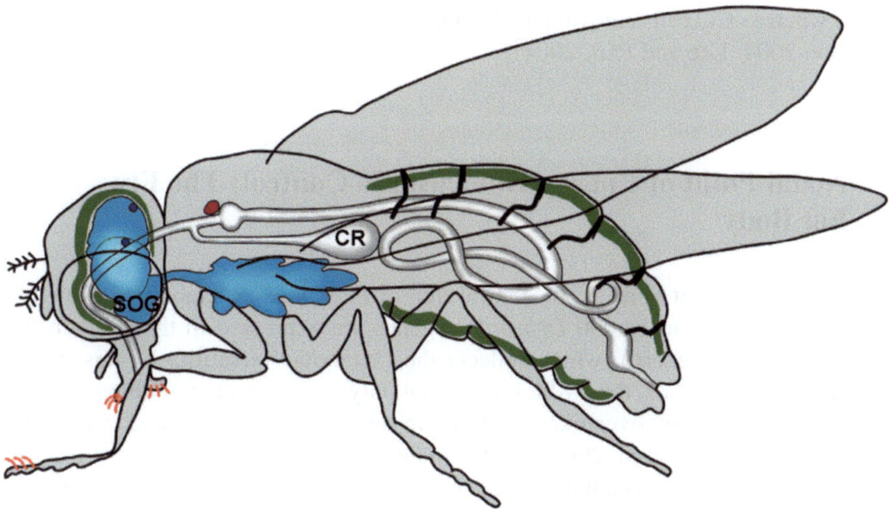


**Fig. 2** The cellular basis of energy homeostasis regulation in the fly. Chemosensory sensilla in the mouthparts and legs (*orange*) initiate the feeding behaviour likely coordinated by the suboesophageal ganglion (SOG) of the central nervous system (*blue*). Ingested food is transiently stored in the crop (CR) of the alimentary canal (*grey*) prior to digestion and nutrient absorption. Energy storage and mobilization is governed by the fat body (*green*) supported by the oenocytes (*black*). Humeral control of the brain-endocrine axis largely resides in the insulin producing cells (*dark blue*) of the central brain and the corpora cardiaca cells of the ring gland (*red*)

of the midgut epithelium suggests resynthesis and transient storage of the dietary lipid components prior to transport to the fat body, which serves as metabolic sink for lipid as well as for carbohydrates (see below). Much like in mammals, lipids commute between organs as lipoproteins in the fly body fluid called hemolymph (Pennington and Wells 2002).

Starvation-induced storage fat mobilization depends – at least at the larval stage – on oenocytes, groups of cells located in the cuticle (Fig. 2), which execute some liver-like functions in lipometabolism (Gutierrez et al. 2007). Ablation of adult oenocytes has recently been shown to eliminate cuticular hydrocarbons acting as pheromones (Billeter et al. 2009) but a role of these cells in imaginal storage lipid catabolism remains to be shown.

Of particular interest are cells orchestrating energy homeostasis control by releasing humeral signals. Two prominent cell groups fulfil this function in the fly brain-neuroendocrine system. On the one hand, the insulin-producing cells (IPCs; analogous to mammalian pancreatic β-cells), which are localized in the central brain (Rulifson et al. 2002) and on the other hand, the so-called corpora cardiaca (CC) cells (analogous to mammalian pancreatic α-cells) localized in a neuroendocrine organ named the ring gland (Kim and Rulifson 2004) (Fig. 2). Ablation of IPCs increases blood sugar levels in *Drosophila* larvae (Rulifson et al.

2002) whereas CC cell removal has the opposite effect (Isabel et al. 2005; Kim and Rulifson 2004; Lee and Park 2004).

## 3 Nodal Point of Energy Homeostasis Control: The Fly Fat Body

Arguably, the fly fat body is the central hub of *Drosophila* metabolism. Roughly half of the more than 14,000 *Drosophila* genes are expressed in the adult fly fat body (www.Flyatlas.org), which reflects the metabolic versatility of this organ. Compared to mammals, the fly fat body is not only the equivalent of adipose tissue but also serves major liver-like functions in energy homeostasis and xenobiotic detoxification (Yang et al. 2007), as this mammalian organ is absent from insects. Moreover, the fat body is a major organ of the immune system in flies (reviewed in Ferrandon et al. (2007)).

Two ontogenetically independent cell lineages compose the embryonic/larval/pupal fat body (Fig. 3a), and the adult fat body (Fig. 3b), respectively (Hoshizaki

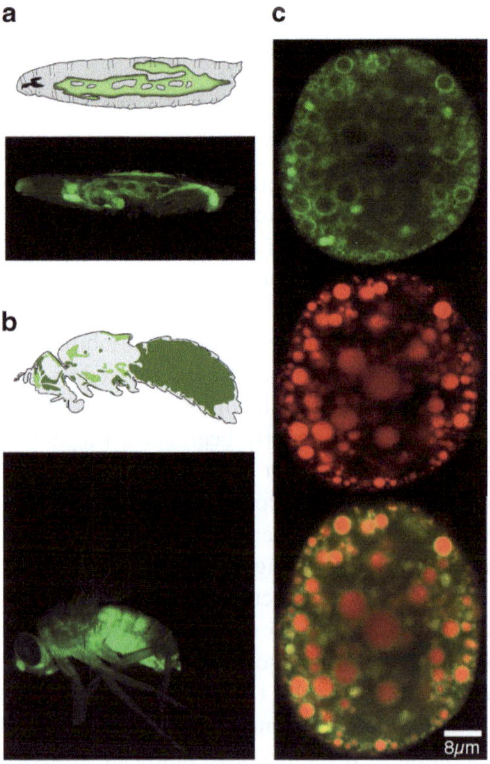


**Fig. 3** Central organ of energy storage – the fat body. Schematic representation and in vivo localization (by targeted GFP expression) of larval (**a**) and adult (**b**) fat body tissue. Note that larval fat body consists of a coherent sheet of cells whereas numerous fat depots make up the adult fat body. Solitary fat body cell (**c**) from an immature adult fly with storage lipid droplets (*red*) decorated by a droplet-associated, GFP-tagged protein (*green*) (for details see text). Note that the photographic picture in (**a**) was originally published in Beller et al. 2006 © The American Society for Biochemistry and Molecular Biology

2005). The larval fat body consists of a defined number of large polyploid cells building a single contiguous organ sheet, which stretches throughout the body cavity and responds to nutrient excess by hypertrophy. In contrast, the adult fat body consists of premitotic cells allocated to various anatomically delimitable fat depots such as the deep visceral or pericardial fat body or the peripheral fat body of head and abdomen (Miller 1950; Fig. 3). Abdominal fat depots of a single fly are estimated to be composed of 18,000 adipocytes (Johnson and Butterworth 1985). However, plasticity of the adult abdominal fat body has recently been demonstrated. This organ responds to insulin signalling by hyperplasia via inhibitory phosphorylation of both, the forkhead box group O (FOXO) transcription factor and glycogen synthase kinase 3 (GSK3) and by hypertrophy via GSK3 inhibition, respectively (Diangelo and Birnbaum 2009). Considering the functional heterogeneity of this organ histological inspection reveals a remarkable uniformity of fat body tissue, which is largely composed of lipid droplet-filled trophocytes at all ontogenetic stages (Miller 1950; Fig. 3c). Beyond doubt, however, the documentation of depot-specific signalling responses (Hwangbo et al. 2004) within the fat body is just an example for a functional complexity, in need to become comprehensively explored.

The fat body is not only the major site of neutral lipid storage in the fly but also governs substantial glycogen stores. Accordingly, the fat body operates as the executive organ of energy storage and mobilization in energy balance control in response to humoral signals of the brain-neuroendocrine system. Recent studies also support feedback regulation of energy homeostasis by the fat body. Organismal growth under starvation is dependent on the upregulation of *Drosophila* insulin-like peptide 6 (DILP6) in the pupal fat body (Slaidina et al. 2009) and the larval fat body remotely controls DILP secretion from the IPCs in the central nervous system by an hitherto unknown humoral signal (Geminard et al. 2009).

## 4 *Drosophila* Lipid Storage Control: More than Fly-Relevant

Interest in the understanding of lipid storage control in the fly has been particularly spurred by two factors: One, by the molecular characterisation of the obese *adipose* mutant fly (Hader et al. 2003), which originated from a Nigerian *Drosophila* wild population some 50 years ago (Doane 1960) and the other, the unanticipated finding that as many as about 70% of human disease genes have bona fide orthologs in the fly genome (Chien et al. 2002; Reiter et al. 2001; www.superfly.ucsd.edu/homophila) enhanced the interest in *Drosophila* lipometabolism research. Subsequent studies identified the mammalian *adipose* ortholog WDTC1 as a novel obesity gene in the mouse (Suh et al. 2007) and man (Lai et al. 2009), providing proof of the concept that fly fat storage regulators are of relevance for the understanding of human lipopathologies.

As far as it has been studied the basic insect lipometabolism (reviewed in Canavoso et al. 2001) largely resembles the corresponding pathways in mammals.

Accordingly, central regulatory mechanisms such as the role of the hepatocyte nuclear factor 4 (HNF4) in lipocatabolism (Palanker et al. 2009) or the phosphorylation of acetyl-*CoA* carboxylase (ACC) by AMP-activated protein kinase (AMPK) (Hardie and Pan 2002) are evolutionarily conserved. So also is the function of sterol regulatory element-binding protein (SREBP) (Kunte et al. 2006; Porstmann et al. 2008) or stearoyl-*CoA* desaturase-1 (SCD-1; called *desat1* in the fly; Ueyama et al. 2005) in lipogenesis. Also *hedgehog* signalling activity suppresses fat storage and/or adipogenesis (Pospisilik et al. 2010; Suh et al. 2006) and the ATP-binding cassette transporter G1 (Buchmann et al. 2007) controls lipid storage in flies and mice alike. On the other hand, important interspecies differences in lipometabolism have been described, such as the feedback regulation of SREBP by sterols in mammals and by phospholipids in flies (Dobrosotskaya et al. 2002).

Exceptional evolutionary conservation has been reported for central components and mechanisms of neutral storage lipid (i.e., triacylglyceride (TAG)) homeostasis. For example organismal lipid understorage has been described for *midway* mutant flies lacking diacylglycerol *O*-acyltransferase 1 (DGAT1) (Buszczak et al. 2002 and Kühnlein unpublished). Conversely, *brummer* mutant flies lacking the *Drosophila* ortholog of mammalian ATGL/PNPLA2 (reviewed in Schweiger et al. 2009) are lipid mobilization-impaired and obese (Grönke et al. 2005) (Fig. 4). These defects are identical to the ATGL loss-of-function phenotype in mice (Haemmerle et al. 2006; Zimmermann et al. 2004) and man (Fischer et al. 2007). In mammals ATGL is activated by CGI-58/ABHD5 in response to β-adrenergic signalling. Signalling triggers protein kinase A (PKA)-dependent phosphorylation of PLIN1/Perilipin and subsequent release of CGI-58 (Lass et al. 2006; Yamaguchi et al. 2007). PLIN1/Perilipin is the founding member of the most prominent lipid droplet-associated protein family called PERILIPINs (Kimmel et al. 2010). Notably, flies do not only encode the ATGL ortholog Brummer but also a CGI-58 homolog and two fly PERILIPINs called PLIN1/LSD-1 and PLIN2/LSD-2 (Fig. 4), all of which localize to lipid droplets (Beller et al. 2006; Cermelli et al. 2006; Grönke et al. 2003). Both fly PERILIPINs control body fat storage in a dosage-dependent manner. PLIN2/LSD-2 mutants are lean, whereas over-expression of the gene causes storage fat accumulation (Grönke et al. 2003), suggesting a barrier function for this protein on the lipid droplet surface. Also PLIN1/LSD-1 can provide a barrier function on lipid droplets of nonadipose tissue. However, PLIN1/LSD-1 mutants are hyperphagic and obese, indicating that the protein facilitates lipid mobilization in the fat body (Beller and Kühnlein unpublished results). Genetic and in vitro evidence support the role of PLIN1/LSD-1 as a downstream effector of the adipokinetic hormone (AKH) signalling pathway (Arrese et al. (2008) and Beller and Kühnlein unpublished results; Fig. 4), one of the two lipolytic pathways in the fly (Grönke et al. 2007). Notably, AKH signalling mediates stimulated lipolysis – like β-adrenergic signalling – via a G protein-coupled receptor (AKHR), cAMP and PKA (Fig. 4). The cAMP regulated transcription factor CREB is another regulatory target of AKH signalling. Expression of dominant negative CREB in the fat body increases fly food intake and causes an obesity-like phenotype (Iijima et al. 2009). Recently, a ceramide synthase encoded by the *schlank* gene, has been demonstrated to

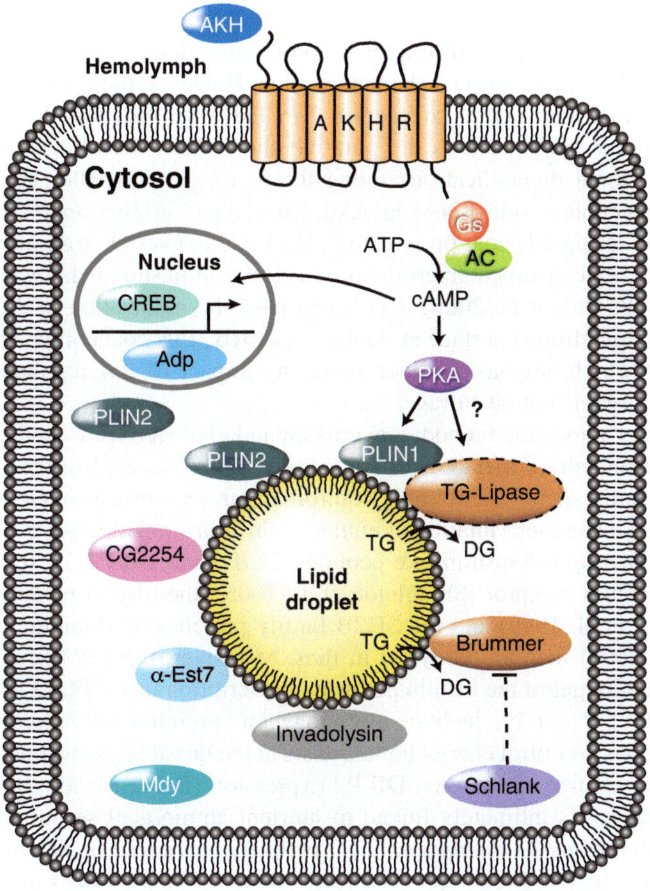


**Fig. 4** Model of lipid storage/mobilization in the fly fat body. The prolipolytic neuropeptide adipokinetic hormone (AKH) binds to its receptor (AKHR) on fat body cells and triggers cAMP second messenger signaling. cAMP activates both, the cAMP responsive element binding protein (CREB) and protein kinase A (PKA). PKA (likely) phosphorylates the PERLIPIN PLIN1/LSD-1, which increases activity of triacylglycerol lipases such as the predicted TG lipase and Brummer. PLIN2/LSD-2 executes barrier function against lipid mobilization. Lipid droplets are decorated by numerous associated proteins of varying distribution and function e.g., Invadolysin, α-Est7 and CG2254 (for details see text). Evolutionarily conserved fly DGAT1 called Midway (Mdy) increases and Adipose (Adp) decreases fat storage. *Arrows* indicate activating and *bar-ending lines* inhibitory interactions. *Dashed* connecting lines represent indirect interactions

profoundly influence neutral lipid storage via transcriptional upregulation of *SREBP* and downregulation of *brummer* (Bauer et al. 2009).

As detailed above, numerous lipometabolism regulators localize to the surface of lipid droplets, which is increasingly acknowledged as the intracellular compartment boundary of central importance for body fat storage. This notion is further supported by the finding that loss of the lipid droplet-associated putative metalloprotease

Invadolysin causes leanness in flies (Cobbe et al. 2009). Many member proteins of the complex *Drosophila* lipid droplet subproteome (Beller et al. 2006; Cermelli et al. 2006) still await functional characterisation. However, the existence of general lipid droplet proteins such as α-Esterase-7 (α-Est7) and proteins decorating lipid droplet subpopulations such as the putative short chain dehydrogenase CG2254 suggest functional diversification among these storage organelles (Beller et al. 2006). Two genome-wide RNAi knockdown screens in *Drosophila* cell culture identified the retrograde coat protein complex I (COPI) vesicle transport system as an evolutionarily conserved determinant of cellular lipid storage homeostasis (Beller et al. 2008; Guo et al. 2008). COPI regulates the association of ATGL to the mammalian lipid droplet surface as well as its PERILIPIN coat composition (Beller et al. 2008) which emphasises once more the impact of this regulatory hub for cellular and organismal fat storage.

As outlined above, the fat body governs fat and glycogen stores and accordingly lipid and carbohydrate metabolism are closely interconnected in the fly. Like in mammals insulin signalling not only controls sugar homeostasis in *Drosophila* but also impacts fat storage. Mutation of numerous insulin signalling pathway members such as the *Drosophila* insulin-like peptides DLIP6 and DILP2-3,5 (Grönke et al. 2010), the insulin receptor (Shingleton et al. 2005), the insulin receptor substrate *chico* (Böhni et al. 1999) and the SH2B family protein *Lnk* (Slaidina et al. 2009; Werz et al. 2009) increase body fat in flies. Moreover, the LIP4 acid lipase is a transcriptional target of the insulin pathway transcription factor FOXO (Vihervaara and Puig 2008). Recently, the bone morphogenetic protein *glass bottom boat* (*gbb*) has been shown to control energy homeostasis in the larval fat body, possibly in part by its remote control of neuronal DILP2 expression (Ballard et al. 2010). Finally, insulin signalling is intimately linked to nutrient amino acid sensing and growth control by the *target of rapamycin* (*Tor*) pathway (reviewed in Teleman (2010)), providing an integrated regulatory network for energy homeostasis in the fly.

## 5 Conclusions and Outlook

The last couple of years witnessed a burst of knowledge about *Drosophila* energy homeostasis. The current picture shows, a remarkable evolutionary conservation between flies and mammals with regard to factors and regulatory mechanisms keeping the organismal energy balance (for an extended view on evolutionarily conserved fly genes affecting lipometabolism see Kühnlein 2010). The comparison of stimulated lipolysis pathways in flies and mammals i.e., AKH and β-adrenergic signalling, respectively, exemplifies the prospects and limitations of the *Drosophila* model system from an anthropocentric view. Whereas the ligands of both pathways differ substantially (i.e., neuropeptide vs. catecholamine) due to ligand/receptor coevolution, factors of the intracellular core-signalling pathway are identical. At the lipid droplet surface, the prolipolytic signal is relayed to PERILIPIN family members in both systems. This protein family consists of only two members in flies but of five in mouse and man, which exemplifies a diversification characteristic for mammalian genomes. Genomic diversification increases regulatory flexibility but functional redundancies among paralogs pose a problem to mutational analysis, less prominent in the fly due to its more simple genomic architecture. Accordingly, *Drosophila* is and will stay a first choice model for functional large-scale approaches to identify energy homeostasis regulators. However, it must fall short of disclosing the functional details of mammalian multigene families nor can it be employed to analyze evolutionary more recent additions to energy homeostasis regulation such as the leptin system.

*Drosophila's* real promise for the energy homeostasis research is the possibility of analyzing the impact of tissue- and time-specific single or combinatorial gene knockdowns in vivo. This approach does justice to the systems biology nature of energy homeostasis and is uniquely capable of modelling polygenic traits, such as many human lipopathologies.

**Acknowledgements** The author is grateful to Herbert Jäckle for continuous support of this research line in his department. Hartmut Sebesse is acknowledged for expert assistance in the illustration layout. This work was supported by the Max Planck Society.

## References

Al-Anzi B, Sapin V, Waters C, Zinn K, Wyman RJ, Benzer S (2009) Obesity-blocking neurons in *Drosophila*. Neuron 63:329–341

Arrese EL, Rivera L, Hamada M, Mirza S, Hartson SD, Weintraub S, Soulages JL (2008) Function and structure of lipid storage droplet protein 1 studied in lipoprotein complexes. Arch Biochem Biophys 473:42–47

Baker KD, Thummel CS (2007) Diabetic larvae and obese flies-emerging studies of metabolism in *Drosophila*. Cell Metab 6:257–266

Ballard SL, Jarolimova J, Wharton KA (2010) Gbb/BMP signaling is required to maintain energy homeostasis in *Drosophila*. Develop Biol 337:375–385

Bass TM, Grandison RC, Wong R, Martinez P, Partridge L, Piper MDW (2007) Optimization of dietary restriction protocols in *Drosophila*. J Gerontol A Biol Sci Med Sci 62:1071–1081

Bauer R, Voelzmann A, Breiden B, Schepers U, Farwanah H, Hahn I, Eckardt F, Sandhoff K, Hoch M (2009) Schlank, a member of the ceramide synthase family controls growth and body fat in *Drosophila*. EMBO J 28:3706–3716

Belay AT, Scheiner R, So AK-C, Douglas SJ, Chakaborty-Chatterjee M, Levine JD, Sokolowski MB (2007) The foraging gene of *Drosophila melanogaster*: spatial-expression analysis and sucrose responsiveness. J Comp Neurol 504:570–582

Beller M, Riedel D, Jänsch L, Dieterich G, Wehland J, Jäckle H, Kühnlein RP (2006) Characterization of the *Drosophila* lipid droplet subproteome. Mol Cell Proteomics 5:1082–1094

Beller M, Sztalryd C, Southall N, Bell M, Jackle H, Auld DS, Oliver B (2008) COPI complex is a regulator of lipid homeostasis. PLoS Biol 6:e292

Billeter J-C, Atallah J, Krupp JJ, Millar JG, Levine JD (2009) Specialized cells tag sexual and species identity in *Drosophila melanogaster*. Nature 461:987–991

Böhni R, Riesgo-Escovar J, Oldham S, Brogiolo W, Stocker H, Andruss BF, Beckingham K, Hafen E (1999) Autonomous control of cell and organ size by CHICO, a *Drosophila* homolog of vertebrate IRS1-4. Cell 97:865–875

Buch S, Pankratz MJ (2009) Making metabolic decisions in *Drosophila*. Fly 3:74–77

Buchmann J, Meyer C, Neschen S, Augustin R, Schmolz K, Kluge R, Al-Hasani H, Jurgens H, Eulenberg K, Wehr R et al (2007) Ablation of the cholesterol transporter adenosine triphosphate-binding cassette transporter G1 reduces adipose cell size and protects against diet-induced obesity. Endocrinology 148:1561–1573

Buszczak M, Lu X, Segraves WA, Chang TY, Cooley L (2002) Mutations in the midway gene disrupt a *Drosophila* acyl coenzyme A: diacylglycerol acyltransferase. Genetics 160: 1511–1518

Canavoso LE, Jouni ZE, Karnas KJ, Pennington JE, Wells MA (2001) Fat metabolism in insects. Annu Rev Nutr 21:23–46

Cermelli S, Guo Y, Gross SP, Welte MA (2006) The lipid-droplet proteome reveals that droplets are a protein-storage depot. Curr Biol 16:1783–1795

Chatterjee A, Tanoue S, Houl JH, Hardin PE (2010) Regulation of gustatory physiology and appetitive behavior by the *Drosophila* circadian clock. Curr Biol 20:300–309

Chien S, Reiter LT, Bier E, Gribskov M (2002) Homophila: human disease gene cognates in *Drosophila*. Nucleic Acids Res 30:149–151

Cobbe N, Marshall K, Rao S, Chang C, Di Cara F, Duca E, Vass S, Kassan A, Heck M (2009) The conserved metalloprotease invadolysin localizes to the surface of lipid droplets. J Cell Sci 122:3414–3423

Diangelo J, Birnbaum M (2009) The regulation of fat cell mass by insulin in *Drosophila melanogaster*. Mol Cell Biol 29:6341–6352

Doane WW (1960) Developmental physiology of the mutant female sterile(2)adipose of *Drosophila melanogaster*. I. Adult morphology, longevity, egg production, and egg lethality. J Exp Zool 145:1–21

Dobrosotskaya IY, Seegmiller AC, Brown MS, Goldstein JL, Rawson RB (2002) Regulation of SREBP processing and membrane lipid production by phospholipids in *Drosophila*. Science 296:879–883

Ferrandon D, Imler J-L, Hetru C, Hoffmann JA (2007) The *Drosophila* systemic immune response: sensing and signalling during bacterial and fungal infections. Nat Rev Immunol 7:862–874

Fischer J, Lefevre C, Morava E, Mussini JM, Laforet P, Negre-Salvayre A, Lathrop M, Salvayre R (2007) The gene encoding adipose triglyceride lipase (PNPLA2) is mutated in neutral lipid storage disease with myopathy. Nat Genet 39:28–30

Geminard C, Rulifson EJ, Leopold P (2009) Remote control of insulin secretion by fat cells in *Drosophila*. Cell Metab 10:199–207

Gordon MD, Scott K (2009) Motor control in a *Drosophila* taste circuit. Neuron 61:373–384

Grandison R, Piper M, Partridge L (2009). Amino-acid imbalance explains extension of lifespan by dietary restriction in *Drosophila*. Nature 462:1061–1064

Grönke S, Beller M, Fellert S, Ramakrishnan H, Jäckle H, Kühnlein RP (2003) Control of fat storage by a *Drosophila* PAT domain protein. Curr Biol 13:603–606

Grönke S, Mildner A, Fellert S, Tennagels N, Petry S, Müller G, Jäckle H, Kühnlein R (2005) Brummer lipase is an evolutionary conserved fat storage regulator in *Drosophila*. Cell Metab 1:323–330

Grönke S, Müller G, Hirsch J, Fellert S, Andreou A, Haase T, Jäckle H, Kühnlein R (2007) Dual lipolytic control of body fat storage and mobilization in *Drosophila*. PLoS Biol 5:e137

Grönke S, Clarke D-F, Broughton S, Andrews TD, Partridge L (2010) Molecular evolution and functional characterization of *Drosophila* insulin-like peptides. PLoS Genet 6:e1000857

Guo Y, Walther TC, Rao M, Stuurman N, Goshima G, Terayama K, Wong JS, Vale RD, Walter P, Farese RV (2008) Functional genomic screen reveals genes involved in lipid-droplet formation and utilization. Nature 453:657–661

Gutierrez E, Wiggins D, Fielding B, Gould AP (2007) Specialized hepatocyte-like cells regulate *Drosophila* lipid metabolism. Nature 445:275–280

Hader T, Muller S, Aguilera M, Eulenberg K, Steuernagel A, Ciossek T, Kuhnlein R, Lemaire L, Fritsch R, Dohrmann C et al (2003) Control of triglyceride storage by a WD40/TPR-domain protein. EMBO Rep 4:511–516

Haemmerle G, Lass A, Zimmermann R, Gorkiewicz G, Meyer C, Rozman J, Heldmaier G, Maier R, Theussl C, Eder S et al (2006) Defective lipolysis and altered energy metabolism in mice lacking adipose triglyceride lipase. Science 312:734–737

Hardie DG, Pan DA (2002) Regulation of fatty acid synthesis and oxidation by the AMP-activated protein kinase. Biochem Soc Trans 30:1064–1070

Hoshizaki DK (2005). Fat-cell development. Comprehensive molecular insect science 2:315–345

Hwangbo D, Gersham B, Tu M, Palmer M, Tatar M (2004) *Drosophila* dFOXO controls lifespan and regulates insulin signalling in brain and fat body. Nature 429:562–566

Iijima K, Zhao L, Shenton C, Iijima-Ando K (2009) Regulation of energy stores and feeding by neuronal and peripheral CREB activity in *Drosophila*. PLoS One 4:e8498

Isabel G, Martin JR, Chidami S, Veenstra JA, Rosay P (2005) AKH-producing neuroendocrine cell ablation decreases trehalose and induces behavioral changes in *Drosophila*. Am J Physiol Regul Integr Comp Physiol 288:R531–538

Ja W, Carvalho G, Mak E, de la Rosa N, Fang A, Liong J, Brummel T, Benzer S (2007) Prandiology of *Drosophila* and the CAFE assay. Proc Natl Acad Sci USA 104:8253–8256

Johnson MB, Butterworth FM (1985) Maturation and aging of adult fat body and oenocytes in *Drosophila* as revealed by light microscopic morphometry. J Morphol 184:51–59

Kamleh MA, Hobani Y, Dow JAT, Zheng L, Watson DG (2009) Towards a platform for the metabonomic profiling of different strains of *Drosophila melanogaster* using liquid chromatography-Fourier transform mass spectrometry. FEBS J 276:6798–6809

Kim S, Rulifson E (2004) Conserved mechanisms of glucose sensing and regulation by *Drosophila* corpora cardiaca cells. Nature 431:316–320

Kimmel AR, Brasaemle DL, McAndrews-Hill M, Sztalryd C, Londos C (2010) Adoption of PERILIPIN as a unifying nomenclature for the mammalian PAT-family of intracellular lipid storage droplet proteins. J Lipid Res 51:468–471

Kühnlein RP (2010) *Drosophila* as a lipotoxicity model organism – more than a promise? Biochim Biophys Acta 1801:215–221

Kunte A, Matthews K, Rawson R (2006) Fatty acid auxotrophy in *Drosophila* larvae lacking SREBP. Cell Metab 3:439–448

Lai CQ, Parnell LD, Arnett DK, Garcia-Bailo B, Tsai MY, Kabagambe EK, Straka RJ, Province MA, An P, Borecki IB et al (2009) WDTC1, the ortholog of *Drosophila* adipose gene, associates with human obesity, modulated by MUFA intake. Obesity (Silver Spring) 17:593–600

Lass A, Zimmermann R, Haemmerle G, Riederer M, Schoiswohl G, Schweiger M, Kienesberger P, Strauss JG, Gorkiewicz G, Zechner R (2006) Adipose triglyceride lipase-mediated lipolysis of cellular fat stores is activated by CGI-58 and defective in Chanarin-Dorfman Syndrome. Cell Metab 3:309–319

Lee G, Park J (2004) Hemolymph sugar homeostasis and starvation-induced hyperactivity affected by genetic manipulations of the adipokinetic hormone-encoding gene in *Drosophila melanogaster*. Genetics 167:311–323

Leopold P, Perrimon N (2007) *Drosophila* and the genetics of the internal milieu. Nature 450:186–188

Ma R, Taruttis A, Ntziachristos V, Razansky D (2009) Multispectral optoacoustic tomography (MSOT) scanner for whole-body small animal imaging. Opt Express 17:21414–21426

McGuire SE, Roman G, Davis RL (2004) Gene expression systems in *Drosophila*: a synthesis of time and space. Trends Genet 20:384–391

Melcher C, Bader R, Pankratz MJ (2007) Amino acids, taste circuits, and feeding behavior in *Drosophila*: towards understanding the psychology of feeding in flies and man. J Endocr 192:467–472

Meunier N, Belgacem YH, Martin J-R (2007) Regulation of feeding behaviour and locomotor activity by takeout in *Drosophila*. J Exp Biol 210:1424–1434

Miller A (1950) The internal anatomy and histology of the imago of *Drosophila melanogaster*. In: Demerec M (ed) Biology of *Drosophila*. Cold Spring Harbour Laboratory, New York, pp 420–534

Null B, Liu CW, Hedehus M, Conolly S, Davis RW (2008) High-resolution, in vivo magnetic resonance imaging of *Drosophila* at 18.8 Tesla. PLoS One 3:e2817

Palanker L, Tennessen JM, Lam G, Thummel CS (2009) *Drosophila* HNF4 regulates lipid mobilization and beta-oxidation. Cell Metab 9:228–239

Pennington J, Wells M (2002) Triacylglycerol-rich lipophorins are found in the dipteran infraorder Culicomorpha, not just in mosquitoes. J Insect Sci 2:15

Porstmann T, Santos CR, Griffiths B, Cully M, Wu M, Leevers S, Griffiths JR, Chung Y-L, Schulze A (2008) SREBP activity is regulated by mTORC1 and contributes to Akt-dependent cell growth. Cell Metab 8:224–236

Pospisilik JA, Schramek D, Schnidar H, Cronin SJF, Nehme NT, Zhang X, Knauf C, Cani PD, Aumayr K, Todoric J et al (2010) *Drosophila* genome-wide obesity screen reveals hedgehog as a determinant of brown versus white adipose cell fate. Cell 140:148–160

Reiter LT, Potocki L, Chien S, Gribskov M, Bier E (2001) A systematic analysis of human disease-associated gene sequences in *Drosophila melanogaster*. Genome Res 11:1114–1125

Righi V, Apidianakis Y, Rahme LG, Tzika AA (2010). Magnetic resonance spectroscopy of live *Drosophila melanogaster* using magic angle spinning. J Vis Exp 38. http://www.jove.com/index/details.stp?id=1710, doi: 10.3791/1710

Rulifson EJ, Kim SK, Nusse R (2002) Ablation of insulin-producing neurons in flies: growth and diabetic phenotypes. Science 296:1118–1120

Schweiger M, Lass A, Zimmermann R, Eichmann TO, Zechner R (2009) Neutral lipid storage disease: genetic disorders caused by mutations in adipose triglyceride lipase/PNPLA2 or CGI-58/ABHD5. Am J Physiol Endocrinol Metab 297:E289–E296

Shingleton AW, Das J, Vinicius L, Stern DL (2005) The temporal requirements for insulin signaling during development in *Drosophila*. PLoS Biol 3:e289

Sieber MH, Thummel CS (2009) The DHR96 nuclear receptor controls triacylglycerol homeostasis in *Drosophila*. Cell Metab 10:481–490

Slaidina M, Delanoue R, Gronke S, Partridge L, Léopold P (2009) A *Drosophila* insulin-like peptide promotes growth during nonfeeding states. Develop Cell 17:874–884

St Johnston D (2002) The art and design of genetic screens: *Drosophila melanogaster*. Nat Rev Genet 3:176–188

Suh J, Gao X, McKay J, McKay R, Salo Z, Graff J (2006) Hedgehog signaling plays a conserved role in inhibiting fat formation. Cell Metab 3:25–34

Suh J, Zeve D, McKay R, Seo J, Salo Z, Li R, Wang M, Graff J (2007) Adipose is a conserved dosage-sensitive antiobesity gene. Cell Metab 6:195–207

Teleman AA (2010) Molecular mechanisms of metabolic regulation by insulin in *Drosophila*. Biochem J 425:13–26

Tweedie S, Ashburner M, Falls K, Leyland P, McQuilton P, Marygold S, Millburn G, Osumi-Sutherland D, Schroeder A, Seal R et al (2009) FlyBase: enhancing *Drosophila* gene ontology annotations. Nucleic Acids Res 37:D555–559

Ueyama M, Chertemps T, Labeur C, Wicker-Thomas C (2005) Mutations in the desat1 gene reduces the production of courtship stimulatory pheromones through a marked effect on fatty acids in *Drosophila melanogaster*. Insect Biochem Mol Biol 35:911–920

Van Voorhies WA, Melvin RG, Ballard JWO, Williams JB (2008) Validation of manometric microrespirometers for measuring oxygen consumption in small arthropods. J Insect Physiol 54:1132–1137

Vihervaara T, Puig O (2008) dFOXO regulates transcription of a *Drosophila* acid lipase. J Mol Biol 376:1215–1223

Vinegoni C, Pitsouli C, Razansky D, Perrimon N, Ntziachristos V (2008) In vivo imaging of *Drosophila melanogaster* pupae with mesoscopic fluorescence tomography. Nat Methods 5:45–47

Werz C, Köhler K, Hafen E, Stocker H (2009) The *Drosophila* SH2B family adaptor Lnk acts in parallel to chico in the insulin signaling pathway. PLoS Genet 5:e1000596
Wong R, Piper MDW, Wertheim B, Partridge L (2009) Quantification of food intake in *Drosophila*. PLoS One 4:e6063
Wu JS, Luo L (2006) A protocol for mosaic analysis with a repressible cell marker (MARCM) in *Drosophila*. Nat Protoc 1:2583–2589
Xu K, Zheng X, Sehgal A (2008) Regulation of feeding and metabolism by neuronal and peripheral clocks in *Drosophila*. Cell Metab 8:289–300
Yamaguchi T, Omatsu N, Morimoto E, Nakashima H, Ueno K, Tanaka T, Satouchi K, Hirose F, Osumi T (2007) CGI-58 facilitates lipolysis on lipid droplets but is not involved in the vesiculation of lipid droplets caused by hormonal stimulation. J Lipid Res 48:1078–1089
Yang J, McCart C, Woods DJ, Terhzaz S, Greenwood KG, Ffrench-Constant RH, Dow JAT (2007) A *Drosophila* systems approach to xenobiotic metabolism. Physiol Genomics 30:223–231
Zimmermann R, Strauss JG, Haemmerle G, Schoiswohl G, Birner-Gruenberger R, Riederer M, Lass A, Neuberger G, Eisenhaber F, Hermetter A, Zechner R (2004) Fat mobilization in adipose tissue is promoted by adipose triglyceride lipase. Science 306:1383–1386
Zinke I, Schütz CS, Katzenberger JD, Bauer M, Pankratz MJ (2002) Nutrient control of gene expression in *Drosophila*: microarray analysis of starvation and sugar-dependent response. EMBO J 21:6162–6173

# Towards Understanding Regulation of Energy Homeostasis by Ceramide Synthases

**Reinhard Bauer**

**Abstract** Energy homeostasis and growth require the coordinated regulation of lipid metabolism. The underlying molecular mechanisms are poorly understood. We are interested in identifying key regulators of lipid homeostasis and their functional mechanism. Recently, we identified the *schlank* gene as a major regulator of lipid homeostasis in *Drosophila*. *Schlank* encodes a conserved member of the Lass/CerS family of ceramide synthases, which contain a catalytic Lag1 motif and a homeobox transcription factor domain. *Schlank* mutant larvae, show decreased levels of sphingolipids and depleted fat stores due to an upregulation of triacylglycerol lipases and a downregulation of SREBP-dependent fatty acid synthesis. In addition, we have demonstrated that mammalian members of the conserved Lass/CerS family had also effects on lipid homeostasis. Therefore, we are currently interested to find how members of this family e.g., *schlank* may act as regulators coordinating cellular and organismic lipid homeostasis in animals mechanistically. We now address these issues by using a combination of genetics, biochemistry and integrative physiology.

## 1 Introduction

Energy homeostasis is critical for normal growth and development (Hay and Sonenberg 2004; Ellisen 2005). It depends, on the ability to control storage and mobilization of fat reserves, during times of energy deprivation such as fasting and exercise. Storage fat is deposited as triglycerides, also called triacylglycerols (TAG), in intracellular lipid droplets which accumulate in specialized organs such

---

R. Bauer
Life & Medical Sciences Institute (LIMES), Program Unit Development, Genetics & Molecular Physiology, Rheinische Friedrich-Wilhelms-University Bonn, Carl-Troll-Str. 31, 53115 Bonn, Germany
e-mail: r.bauer@uni-bonn.de

W. Meyerhof et al. (eds.), *Sensory and Metabolic Control of Energy Balance*, Results and Problems in Cell Differentiation 52,
DOI 10.1007/978-3-642-14426-4_14, 

as mammalian adipose tissue or the fat body of flies. In mammals and insects, starvation-induced lipolysis of TAGs is induced by lipases and leads to the release of fatty acids and other lipids into the circulation (Arquier and Leopold 2007 and references herein). The lipids are then taken up and broken down by hepatocytes in the mammalian liver or oenocytes in *Drosophila* (Gutierrez et al. 2007). In the fed state, hepatocytes and oenocytes are also able to synthesize fatty acids for incorporation into TAGs. These can then be assembled into lipoprotein particles, delivered to adipocytes or the insect fat body cells and stored in lipid droplets. Chronic dysregulation of the balance between lipolysis and lipogenesis may lead to metabolic abnormalities such as obesity, lipodystrophy syndromes (Simha and Garg 2006) or insulin resistance in humans.

During energy consuming activities such as growth and reproduction, the coordinated regulation of the lipid metabolism is important on both the cellular and systemic level. This coordination is described by two paradigms: sensing of energy levels mediated by AMPK signaling (Hietakangas and Cohen 2009, review) and sterol/lipid sensing by SREBPs (sterol element binding proteins), which are membrane bound transcription factors that monitor cell membrane composition and adjust lipid synthesis accordingly (Kunte et al. 2006; Dobrosotskaya et al. 2002). Equally important is *de novo* synthesis of sphingolipids. They are essential structural components of eukaryotic membranes and play important roles as second messengers regulating apoptosis, survival and differentiation (Pettus et al. 2002; Levy and Futerman 2010). Enzymes of the sphingolipid pathway are well conserved in all genetically studied eukaryotes (Venkataraman and Futerman 2002) and misregulation of the sphingolipid metabolism is involved in the cause and pathology of several human diseases, including neurodegeneration, cancer, immunity and diabetes (Lahiri and Futerman 2007). On the organismic level, lipid and carbohydrate metabolism are regulated by a variety of hormones, including insulin and glucagon in mammals or insulin-like peptides and the glucagon-like adipokinetic hormones (Akhs) in insects (Van der Horst et al. 2001; Brogiolo et al. 2001). While insulin-like peptides promote cellular glucose import and energy storage in the form of glycogen and TAGs, glucagon or Akh increase lipolysis and glycogenolysis during energy-requiring activities (Van der Horst et al. 2001; Brogiolo et al. 2001; Hafen 2004). Mechanisms by which cellular lipid metabolism is interlaced with hormone-dependent body fat regulation are mostly unknown at present.

In order to identify regulators of energy and lipid homeostasis and to investigate their function, my group is also resorting to taking several approaches.

## 2 *Schlank* Controls Growth and Body Fat

To isolate novel regulators we use various genetic tools like P-element mutation and RNAi-mediated *knockdown* for a genetic screen. Additionally, gene targeting is used to introduce specific mutations in selected genes to study their function in detail. As energy homeostasis is critical for normal growth and development, we screened for genes controlling larval growth in *Drosophila*. This can easily be scored under the dissecting microscope. Here, I will focus on a mutant generated by transposon insertions in a novel locus on the X-chromosome, which we named *schlank*. The *schlank* gene encodes a protein containing six putative transmembrane (TM) domains, a Lag1 motif and a homeobox (Hox) domain (Fig. 1). It shows homology to members of the highly conserved Longevity assurance homologue (Lass) family of ceramide synthases (CerS). The homologies range from yeast to mammals, which have six paralogs. All Lass/CerS genes encode a highly conserved Lag1 motif, which is functionally required for ceramide synthesis and most of them have an additional Hox domain. For all Lass/CerS proteins analyzed so far a function as ceramide synthase was shown. Ceramide biosynthesis plays a key role in the sphingolipid biosynthetic pathway. Phenotype analysis of *schlank* mutants indicated an essential role in regulating larval growth. In addition, we observed that some of our mutants appeared much slimmer and somewhat transparent as compared to control animals hinting to reduced, body fat stores (Fig. 2).

To exclude, that the growth or "thick-thin" phenotypes are just caused by starvation, we routinely perform an assay feeding red coloured yeast. It shows whether the larvae are taking up food or not. Second, growth phenotypes caused by a mutation in an enzyme-encoding gene require the proof of enzymatic activity. Therefore, in collaboration with the Sandhoff group (Bonn), we developed an enzymatic assay to test the activity of the putative ceramide synthase *schlank*. It combines feeding assays and lipid analysis. Here, we offer radiolabelled L-[3-$^{14}$C]-serine, a precursor of sphingolipid biosynthesis followed by the extraction of lipids and ceramides, which are then separated by Thin Layer Chromatography (TLC) and further analyzed by Mass Spectrometry. This system allows us determine and quantify *de novo* generated ceramides. The versatility of this assay system may be expanded by the combination with *Drosophila* genetic

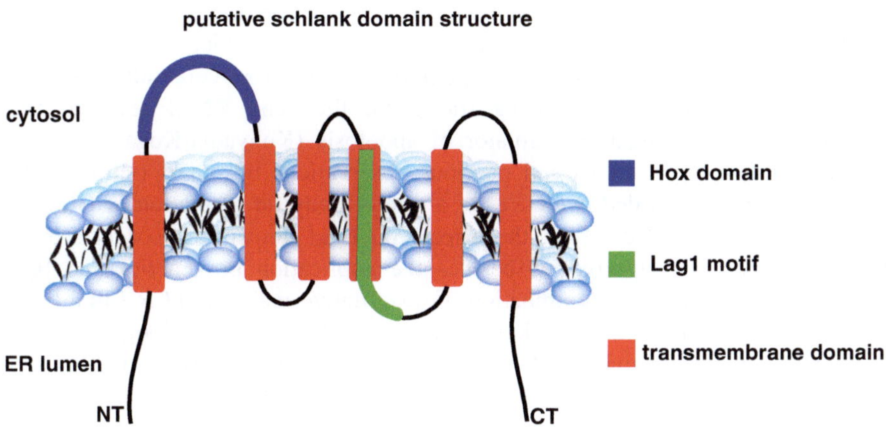


**Fig. 1** Putative domain structure of *Drosophila* ceramide synthase schlank (modified according Bauer et al. 2009)

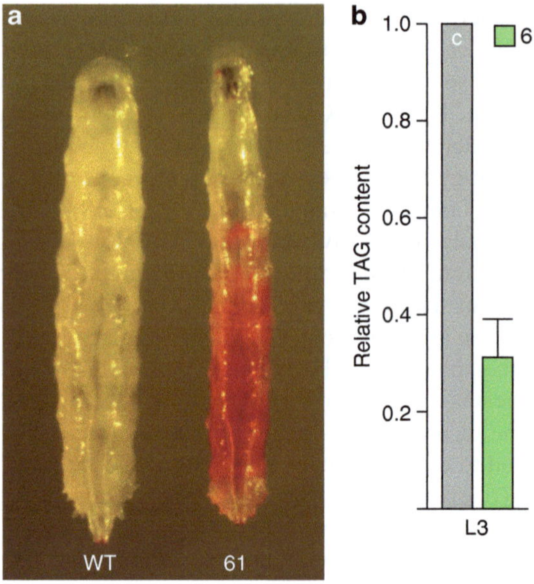


**Fig. 2** (**a**) Mutants larvae carrying the *schlank*$^{G0061}$ (61) allele (Bauer et al. 2009) are thinner than wildtype (WT) larvae when reaching the same size at L3 stage. (**b**) Triacylglycerol (TAG) level of *schlank*$^{G0061}$ (*green bar*) compared to WT in third instar wandering stage larvae

tools such as overexpression or downregulation of the gene of interest using the binary Gal4/UAS system (Brand and Perrimon 1993). In this way we demonstrated the function of *schlank* as a ceramide synthase, as we expected. This assay system enabled us also to prove that the thin looking *schlank* mutants had indeed reduced body fat stores. Applying *schlank* RNAi or overexpressing *schlank* in *Drosophila* larvae led either to emptied or replenished body fat stores (Bauer et al. 2009). Noteworthy, overexpression of the mouse Lass2/CerS2 protein in *Drosophila* larvae resulted in a similar increase of the larvae's body fat. With a similar experimental approach we are currently investigating glucosylceramide synthase (GlcT-1), which catalyzes the formation of glucosylceramide (GlcCer), the core structure of major glycosphingolipids (GSLs), from ceramide and UDP-glucose. Functional analysis using transgenic flies and RNAi revealed that DGlcT-1 acts as a negative regulator of apoptosis (Kohyama-Koganeya et al. 2004). Our first experiments and previous observations by others (Hirabayashi, Japan) have now provided evidence for a role in body fat regulation of GlcT-1.

In order to understand how the effects on lipid homeostasis occur we are following additional approaches besides the application of genetics and lipid biochemistry. We use the combination of quantitative real time PCR (qRT PCR) and immunoblotting to analyze key regulators of lipolysis and lipogenesis. We found that in *schlank* mutants, mRNA levels of lipases (e.g., *brummer* and *lipase3*) were strongly increased, whereas the expression of SREBP was significantly decreased (Bauer et al. 2009). With these data it now appears that reduced body fat e.g., TAGs are not only a result of, increased lipolysis, but also of decreased lipogenesis. Together with data from other labs showing that modulation of ceramide synthesis can add to SREBP regulation, it becomes evident that the observed reduced ceramide synthase activity and reduced lipogenesis in *schlank* mutants may be mediated at least in part by downregulation of SREBP. Taken together these data challenge the view on ceramide synthases. During the process of ceramide synthesis fatty acids are required twice. First, at the initiation step of *de novo* sphingolipid synthesis, the condensation reaction of serine and a long-chain fatty acid lead to 3-ketosphinganine. Second, during the transfer of fatty acids to (dihydro)sphingosine by aminoacylation of sphingosine carbon-2 with a long chain fatty acid yielding dihydroceramide, which is further converted to ceramide. Thus, we think that *schlank* might act as a metabolic sensor, which coordinates sphingolipid metabolism with fatty acid metabolism. A role in regulating organismal fat storage or mobilization has previously not been observed for Lass/CerS family members due to the lack of animal models.

## 3 Outlook

Questions arising now are: how can *schlank* mediate its effect on lipolysis and lipogenesis mechanistically and how is *schlank* expression regulated? Over the past couple of years, many groups have shown a role for ceramide, the building block of all sphingolipids, as an important intracellular signalling molecule involved in regulating differentiation, proliferation, and apoptosis (Pettus et al. 2002; Levy and Futerman 2010). Additional roles of ceramide biosynthesis in insulin resistance and lipotoxicity (Summers 2006), in body weight regulation, energy metabolism, and the metabolic syndrome (Yang et al. 2009) were added.

We recently found that overexpression of a *schlank* variant, which contains a point mutation in the Lag1 motif shown to inhibit ceramide synthase function (Bauer et al. 2009) still caused an increase in TAG comparable to the elevation seen overexpressing, wild type *schlank* or murine CerS/Lass2. These data hint towards a mechanism of *schlank* on TAG metabolism independent of its ceramide synthase function. Many of the Lass/CerS proteins of higher organisms including Lass/CerS 2 contain a *N*-terminally located Hox domain. It is an attractive idea to speculate that this domain may be involved in some of the regulatory effects exerted by *schlank*.

In our lab we are currently applying a genomic engineering approach to address this question.

Genomic engineering permits directed and highly efficient modifications of a chosen genomic locus into virtually any desired mutant allele (Huang et al. 2009). Using this approach for *schlank*'s native locus, we will introduce a set of defined mutant alleles that are strategically designed to test hypotheses about *schlank*'s domains *in vivo* functions and interactions.

Lipid and carbohydrate metabolism are regulated by a variety of hormones, including insulin and glucagon in mammals or insulin-like peptides and the glucagon-like adipokinetic hormones in insects (Brogiolo et al. 2001; Van der

Horst et al. 2001; Fuss et al. 2006). However, mechanisms by which cellular lipid metabolism might be interlaced with hormone-dependent body fat regulation are unknown at present. In our lab Voelzmann succeeded in identifying several *schlank* regulatory regions. In combination with reporter assay systems both, in *Drosophila* animals or in tissue culture, the influence of metabolic signalling cascades on *schlank* regulation, such as insulin signalling, can now be studied in more detail.

# 4 Conclusion

We have recently identified the *Drosophila* ceramide synthase *schlank* as a new regulator of growth and lipid homeostasis (Bauer et al. 2009). Besides the role of schlank in ceramide synthesis we found that *schlank* can influence body fat metabolism by regulating the equilibrium between lipogenesis and lipolysis during larval growth. Our current and future functional analysis of the schlank ceramide synthase will significantly contribute to answer the question about the regulation of Lass/CerS proteins transcriptionally and post transcriptionally in invertebrates and mammals. In turn, we will also gain information on the regulatory role in energy metabolism by schlank. A major challenge ahead will be to evaluate and compare the data that will be obtained from the *Drosophila* in vivo model with the mouse *in vivo* model.

# References


Arquier N, Léopold P (2007) Fly foie gras: modeling fatty liver in Drosophila. Cell Metab 5:83–85

Bauer R, Voelzmann A, Breiden B, Schepers U, Farwanah H, Hahn I, Eckardt F, Sandhoff K, Hoch M (2009) Schlank, a member of the ceramide synthase family controls growth and body fat in *Drosophila*. EMBO J 28:3706–3716

Brand AH, Perrimon N (1993) Targeted gene expression as a means of altering cell fates and generating dominant phenotypes. Development 118:401–415

Brogiolo W, Stocker H, Ikeya T, Rintelen F, Fernandez R, Hafen E (2001) An evolutionarily conserved function of the *Drosophila* insulin receptor and insulin-like peptides in growth control. Curr Biol 11:213–221

Dobrosotskaya IY, Seegmiller AC, Brown MS, Goldstein JL, Rawson RB (2002) Regulation of SREBP processing and membrane lipid production by phospholipids in *Drosophila*. Science 3:879–883

Ellisen LW (2005) Growth control under stress: mTOR regulation through the REDD1-TSC pathway. Cell Cycle 4:1500–1502

Fuss B, Becker T, Zinke I, Hoch M (2006) The cytohesin Steppke is essential for insulin signalling in *Drosophila*. Nature 444:945–948

Gutierrez E, Wiggins D, Fielding B, Gould AP (2007) Specialized hepatocyte-like cells regulate *Drosophila* lipid metabolism. Nature 445:275–280

Hafen E (2004) Cancer, type 2 diabetes, and ageing: news from flies and worms. Swiss Med Wkly 134:711–719

Hay N, Sonenberg N (2004) Upstream and downstream of mTOR. Genes Dev 18:1926–1945

Hietakangas V, Cohen SM (2009) Regulation of tissue growth through nutrient sensing. Annu Rev Genet 43:389–410

Huang J, Zhou W, Dong W, Watson AM, Hong Y (2009) From the cover: directed, efficient, and versatile modifications of the *Drosophila* genome by genomic engineering. Proc Natl Acad Sci USA 106:8284–8249

Kohyama-Koganeya A, Sasamura T, Oshima E, Suzuki E, Nishihara S, Ueda R, Hirabayashi Y (2004) *Drosophila* glucosylceramide synthase: a negative regulator of cell death mediated by proapoptotic factors. J Biol Chem 279:35995–6002

Kunte AS, Matthews KA, Rawson RB (2006) Fatty acid auxotrophy in *Drosophila* larvae lacking SREBP. Cell Metab 3:439–448

Lahiri S, Futerman AH (2007) The metabolism and function of sphingolipids and glycosphingolipids *Cell.* Mol Life Sci 64:2270–2284

Levy M, Futerman AH (2010) Mammalian ceramide synthases. IUBMB Life 62:347–356

Pettus BJ, Chalfant CE, Hannun YA (2002) Ceramide in apoptosis: an overview and current perspectives. Biochim Biophys Acta 1585:114–125

Simha V, Garg A (2006) Lipodystrophy: lessons in lipid and energy metabolism. Curr Opin Lipidol 17:162–169

Summers SA (2006) Ceramides in insulin resistance and lipotoxicity. Prog Lipid Res 45:42–72

Van der Horst DJ, Van Marrewijk WJ, Diederen JH (2001) Adipokinetic hormones of insect: release, signal transduction, and responses. Int Rev Cytol 211:179–240

Venkataraman K, Futerman AH (2002) Do longevity assurance genes containing Hox domains regulate cell development via ceramide synthesis? FEBS Lett 528:3–4

Yang G, Badeanlou L, Bielawski J, Roberts AJ, Hannun YA, Samad F (2009) Central role of ceramide biosynthesis in body weight regulation, energy metabolism, and the metabolic syndrome. Am J Physiol Endocrinol Metab 297:E211–224

# Role of the Gut Peptide Glucose-Induced Insulinomimetic Peptide in Energy Balance

**Andreas F.H. Pfeiffer, Natalia Rudovich, Martin O. Weickert, and Frank Isken**

**Abstract** Glucose-induced insulinomimetic peptide (GIP) is a gut hormone produced by enteroendocrine K-cells in the intestinal mucosa in response to fat, glucose, and also protein. GIP releases insulin from the β cells of the pancreatic islets of Langerhans and therefore is an incretin hormone. GIP acts on a G-protein–coupled receptor that is widely distributed in the body including adipose tissue, stomach, brain, and others. Deletion of the GIP receptor (GIPR) renders mice resistant to weight gain induced by a high fat diet.

We observed that weight gain induced by ovarectomy in female mice is prevented by GIPR deletion that is linked to reduced food intake and reduced hypothalamic expression of orectic neurotransmitters. Moreover, old male $GIPR^{-/-}$ mice placed on a high glycemic index diet maintained a high insulin sensitivity and were much more active than controls, which was not seen in young animals. Thus, GIP elicits central effects in response to nutrients that protect against obesity and insulin resistance. We then investigated the acute responses of humans to treatment with GIP over 4 h in a dose mimicking postprandial plasma levels of about 100 pmol/L. At basal glucose, GIP does not elicit insulin release. Fat biopsies taken before and after 4 h of GIP treatment were analyzed for transcriptomic responses using Agilent whole human genome assays. There was a highly significant upregulation of an inflammatory expression pattern in a pathway analysis.

A.F.H. Pfeiffer (✉), N. Rudovich, M.O. Weickert, and F. Isken
Department of Clinical Nutrition, German Institute of Human Nutrition – Potsdam Rehbrücke, Arthur-Scheunert Allee 114-116, 14558 Nuthetal, Germany
Department of Endocrinology, Diabetes and Nutrition, Charité Universitätsmedizin Berlin, Campus Benjamin Franklin and Campus Charité Mitte, Hindenburgdamm 30, 12200 Berlin, Germany
e-mail: afhp@dife.de

W. Meyerhof et al. (eds.), *Sensory and Metabolic Control of Energy Balance*, Results and Problems in Cell Differentiation 52,
DOI 10.1007/978-3-642-14426-4_15, © Springer-Verlag Berlin Heidelberg 2010

# 1 Introduction

Gut hormones play a central role in energy metabolism. In humans, this is best revealed by the striking success of bariatric surgery for the treatment of morbid obesity. The most successful approach to date among the nonmalabsorptive procedures is the gastric bypass that was shown to cause a sustained reduction in body weight by 20–50 kg even 10 years after the procedure and a successful treatment of type II diabetes and dyslipidemia (Sjostrom et al. 2004). The procedure consists of connecting the upper small bowel, the ileum, to the esophagus. The stomach is closed at the upper end such that no food passes through the stomach, the duodenum, and jejunum and the ileum is reconnected to the small intestine in the form of a "Y," using the strategy named "Roux en Y Gastric Bypass" RYGB. Thereby, nondigested food reaches the ileum and declenches an excessive intestinal hormone response (Ferrannini and Mingrone 2009; Mingrone and Castagneto 2009).

The role of the individual hormones is partly characterized. There is a massively increased release of GLUCAGON-like peptide-1 (GLP-$1_{7\text{-}39}$) and glucose-induced insulinomimetic peptide (GIP) and peptide YY (PYY). GLP-1 and GIP are so-called "incretins" – intestinal hormones released by food components that cause insulin release. Incretins normally account for approximately 60–70% of the insulin response to orally administered glucose but also elicit additional effects on metabolic control. GLP-1 inhibits gastric emptying and reduces glucagon release. GLP-1 and PYY reduce food intake by reducing appetite at central sites. These hormones may be viewed as hormonal systems to protect the organism from excessive food intake.

# 2 Role of GIP in Metabolism Promoting Effective Assimilation and Storage of Food

GIP appears to play a different role in metabolism promoting effective assimilation and storage of food. Remarkably, a genetic deletion of GIPRs (GIPR$^{-/-}$) was shown to protect mice from developing obesity in response to a high fat diet (Miyawaki et al. 2002). GIPR deletion did not alter food intake but increased physical activity in this model. GIPRs are widely distributed in humans and are found not only on β cells of the pancreatic islets of Langerhans but also on fat cells, in gastric mucosa, the lungs, and other organs (Table 1) (Rudovich et al. 2007).

Remarkably, the expression levels of GIPR in humans were higher in abdominal as compared to subcutaneous fat and were highly correlated to many components of the metabolic syndrome such as waist circumference, triglyceride serum levels, HDL-cholesterol, and insulin sensitivity (Rudovich et al. 2007). GIPR expression in subcutaneous adipose tissue was reduced in obese people and increased upon 5% weight loss in a sample of 14 women (Rudovich et al. 2007).

Since genetic deletion of GIPR was shown to prevent high fat diet–induced obesity, we decided to test another model of obesity induced by deficiency of

**Table 1** Correlation of GIP receptor mRNA expression levels determined by qRT-PCR in human adipose tissue biopsies with serum parameters and waist circumference (Rudovich et al. 2007)

| GIPR gene expression in subcutaneous adipose tissue, $n = 95$ | | |
|---|---|---|
| Variable | $r$ | $p$ |
| Age | −0.124 | NS |
| Waist circumference | −0.399 | 0.0001 |
| HOMA-IR | −0.424 | 0.0001 |
| Fasting glucose | −0.400 | 0.0001 |
| Fasting insulin | −0.427 | 0.0001 |
| TG | −0.381 | 0.0001 |
| HDL-cholesterol | 0.393 | 0.0001 |
| LDL-cholesterol | −0.365 | 0.001 |

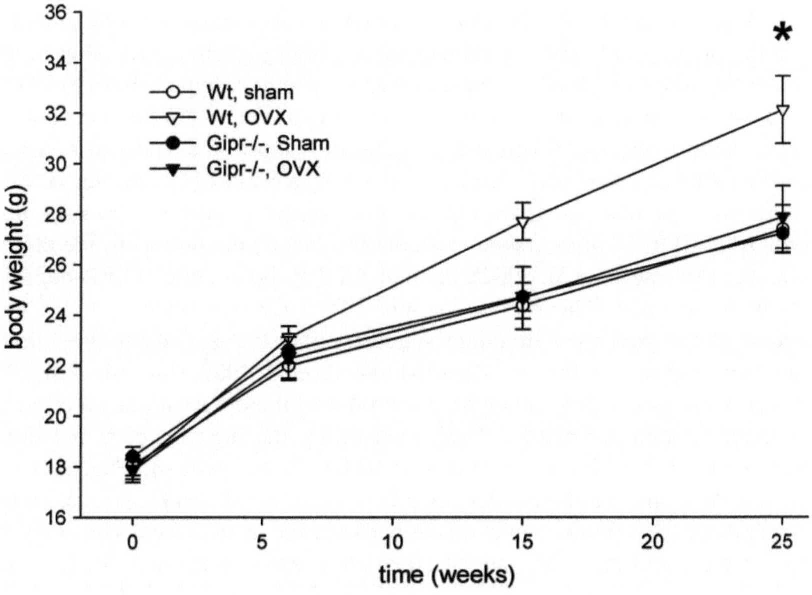


**Fig. 1** Weight gain induced by ovarectomy is prevented by deletion of GIP receptors

estrogens: the ovarectomized mouse (Fig. 1). For this approach, we used the genetic model of $GIPR^{-/-}$ mice. Indeed, after ovarectomy, the mice gained weight on normal laboratory chow compared to sham-operated controls. This weight gain was not observed in the $GIPR^{-/-}$ mice. Ovarectomy did not result in increased food intake, but in a reduction of activity during the night phase, when mice are active normally. The ovarectomy thus reduced nonresting energy expenditure that was, however, also observed both in the control and in the $GIPR^{-/-}$ mice. In order to further elucidate the mechanism, we determined food intake and observed a reduced intake in the $GIPR^{-/-}$ mice compared to controls, which may well explain the increased body weight. It is well known that food intake is influenced by hypothalamic orexigenic and anorexigenic neurohormones.

We therefore determined the mRNA expression of several satiety hormones and observed a reduced expression of the anorexigenic neuropeptide CART (cocaine and amphetamine-related transcript) in both control and GIPR$^{-/-}$ mice. This may well explain the reduced activity in the dark phase. A selective decrease of Neuropeptide Y (NPY) mRNA was only apparent in the GIPR$^{-/-}$ mice but not in controls that we propose to be related to the resistance against obesity upon ovarectomy (Isken et al. 2008). This would suggest that the GIPR influences hypothalamic circuits of neurotransmitters related to appetite although direct effects on appetite have not been described in contrast to GLP-1, which is known to reduce food intake.

GIP is released in the upper intestine in response to carbohydrates, fat, and protein. An inhibition of carbohydrate uptake in the small intestine by an inhibitor of $\alpha$ glycosidase, acarbose, was shown to reduce GIP release very substantially in humans (Enc et al. 2001). We therefore speculated that deletion of GIPR should also prevent the obesity and other consequences of a high glycemic index diet in mice. We first established a high GI diet based on amylopectin as opposed to a low GI diet based on amylose in mice (65% carbohydrate, 23% protein, 12% fat) and confirmed that it produced about twofold greater increases of glucose and insulin in the animals (Isken et al. 2009a). Mice were then fed this diet from 6 weeks of age over 20 weeks, which resulted in obesity. However, similar weight gain was observed in the 6-month-old GIPR$^{-/-}$ mice (Isken et al. 2009b). Thus, in contrast to the effect of high fat diet (Miyawaki et al. 2002), the high GI diet did not elicit GIPR-dependent effects in these mice. The reason for this difference was unclear, but a closer inspection of the published literature suggested that the age of the animals may play a role in response to different diets (Miyawaki et al. 2002; Hansotia et al. 2007). Moreover, work with GIPR antagonists reproduced the observations reported with the genetically modified GIPR$^{-/-}$ mice regarding the high fat diets but did not observe an effect of high carbohydrate diets (Flatt 2008; Irwin and Flatt 2009).

We therefore also studied older mice beginning at 24 weeks of age. In these mice, no difference in body weight developed over the next 20 weeks on either high or low GI diets and there was no difference between control and GIPR$^{-/-}$ mice (Fig. 2). Thus, the age of the animals appears to affect the response to the GI of the diets, and this was independent from the GIPR status. Since more subtle differences might be present in these animals, we undertook a more detailed characterization, including food intake studies and an analysis of activity and insulin sensitivity. Activity was previously shown to be increased in the GIPR$^{-/-}$ mice on high fat diets (Miyawaki et al. 2002; Hansotia et al. 2007) and clearly played an important role in their energy homeostasis.

A further analysis of the role of the GI in the high carbohydrate diet of the animals revealed important differences: the GIPR$^{-/-}$ mice were much more active on the high GI diet than on the low GI diet while such a difference was not observed in the wild-type animals. The GIPR$^{-/-}$ mice on the high GI diet moreover had a significantly higher insulin sensitivity than the wild-type mice on the high GI diet as shown by the decrease in glucose upon i.p. injection of insulin. Moreover, i.p. glucose tolerance tests showed a much better glucose tolerance in the GIPR$^{-/-}$

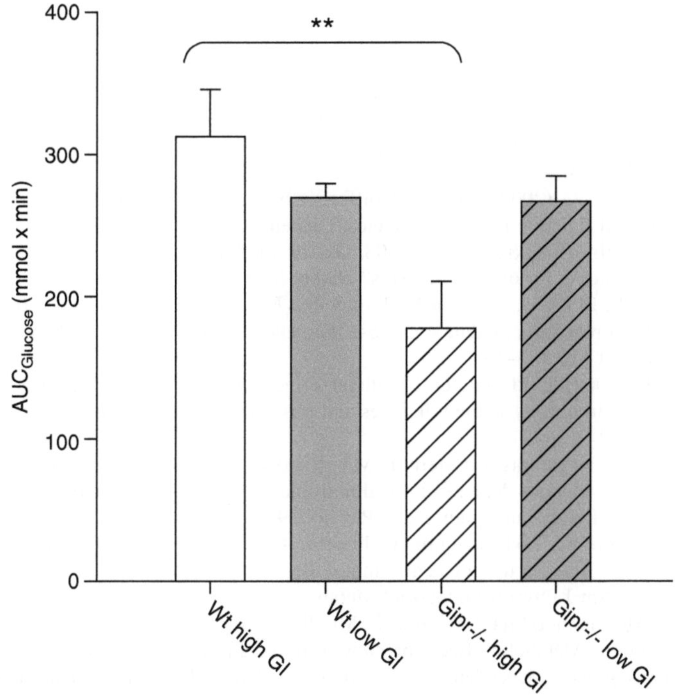


**Fig. 2** Insulin sensitivity is increased in old GIP receptor knock out mice on a high glycemic index diet compared to controls. Insulin tolerance test after i.p. injection of insulin. The *bars* show the area under the curve

mice on the high GI diet. By contrast, the wild-type and the GIPR$^{-/-}$ mice did not differ on the low GI diet. Thus, the high GI diet induced a compensatory increase in physical activity in the GIPR$^{-/-}$ mice, leading to an improvement in insulin sensitivity and glucose tolerance. Since the wild-type mice did not show this compensatory increase in physical activity, it appears that GIP-related signals prevented this behavior (Isken et al. 2009b). This is highly reminiscent of the observations made in high fat diet–fed mice (Miyawaki et al. 2002) and may be interpreted to indicate that GIP is involved in regulating physical activity upon intake of energy-dense diets.

## 3 Conclusion

We conclude that GIP appears to promote a positive energy balance at the central level through different mechanisms depending on the age and sex of the animals.

## References

Enc FY, Imeryuz N, Akin L, Turoglu T, Dede F, Haklar G, Tekesin N, Bekiroglu N, Yegen BC, Rehfeld JF, Holst JJ, Ulusoy NB (2001) Inhibition of gastric emptying by acarbose is correlated with GLP-1 response and accompanied by CCK release. Am J Physiol Gastrointest Liver Physiol 281:G752–763

Ferrannini E, Mingrone G (2009) Impact of different bariatric surgical procedures on insulin action and beta-cell function in type 2 diabetes. Diabetes Care 32:514–520

Flatt PR (2008) Dorothy Hodgkin Lecture 2008. Gastric inhibitory polypeptide (GIP) revisited: a new therapeutic target for obesity-diabetes? Diabet Med 25:759–764

Hansotia T, Maida A, Flock G, Yamada Y, Tsukiyama K, Seino Y, Drucker DJ (2007) Extrapancreatic incretin receptors modulate glucose homeostasis, body weight, and energy expenditure. J Clin Invest 117:143–152

Irwin N, Flatt PR (2009) Evidence for beneficial effects of compromised gastric inhibitory polypeptide action in obesity-related diabetes and possible therapeutic implications. Diabetologia 52:1724–1731

Isken F, Pfeiffer AFH, Nogueiras R, Osterhoff MA, Ristow M, Tschöp M, Weickert MO (2008) Deficiency for the glucose-dependent insulinotropic polypeptide (GIP) receptor prevents ovarectomy-induced body mass gain. Am J Physiol 295:E350–355

Isken F, Klaus S, Petzke KJ, Loddenkemper C, Pfeiffer AF, Weickert MO (2009) Impairment of fat oxidation under high vs low glycemic index diet occurs prior to the development of an obese phenotype. Am J Physiol Endocrinol Metab

Isken F, Weickert MO, Tschop MH, Nogueiras R, Mohlig M, Abdelrahman A, Klaus S, Thorens B, Pfeiffer AF (2009b) Metabolic effects of diets differing in glycaemic index depend on age and endogenous glucose-dependent insulinotrophic polypeptide in mice. Diabetologia 52: 2159–2168

Mingrone G, Castagneto M (2009) Bariatric surgery: unstressing or boosting the beta-cell? Diabetes Obes Metab 11(Suppl 4):130–142

Miyawaki K, Yamada Y, Ban N, Ihara Y, Tsukiyama K, Zhou H, Fujimoto S, Oku A, Tsuda K, Toyokuni S, Hiai H, Mizunoya W, Fushiki T, Holst JJ, Makino M, Tashita A, Kobara Y, Tsubamoto Y, Jinnouchi T, Jomori T, Seino Y (2002) Inhibition of gastric inhibitory polypeptide signaling prevents obesity. Nat Med 8:738–742

Rudovich N, Kaiser S, Engeli S, Osterhoff M, Gogebakan O, Bluher M, Pfeiffer AF (2007) GIP receptor mRNA expression in different fat tissue depots in postmenopausal non-diabetic women. Regul Pept 142:138–145

Sjostrom L, Lindroos AK, Peltonen M, Torgerson J, Bouchard C, Carlsson B, Dahlgren S, Larsson B, Narbro K, Sjostrom CD, Sullivan M, Wedel H (2004) Lifestyle, diabetes, and cardiovascular risk factors 10 years after bariatric surgery. N Engl J Med 351:2683–2693

# Adipocyte–Brain: Crosstalk

**Carla Schulz, Kerstin Paulus, and Hendrik Lehnert**

**Abstract** The initial discovery of leptin, an appetite-suppressing hormone originating from fat tissue, substantially supported the idea that fat-borne factors act on the brain to regulate food intake and energy expenditure. Since then, a growing number of cytokines have been found to be released from adipose tissue, thus acting in an endocrine manner. These adipocytokines include not only, e.g., adiponectin, apelin, resistin, and visfatin, but also inflammatory cytokines and steroid hormones such as estrogens and glucocorticoids. They are secreted from their adipose depots and differ in terms of release stimuli, downstream signaling, and their action on the brain. Clearly, adipocytokines play a prominent role in the central control of body weight, and the deregulation of this circuit may lead to the development of obesity and related disorders. In this chapter, we will focus on crosstalk mechanisms and the deregulation of adipocytokines at the expression level and/or sites of central action that eventually will lead to the development and perpetuation of obesity and diabetes.

## 1 Overview and Discussion

It is known for some time that body energy stores, namely fat, are monitored and tightly regulated. Despite large differences in energy intake and energy expenditure over months and years, body weight remains remarkably stable without any conscious effort. This clearly suggests a system of homeostatic control of body weight.

In the 1940s, Hetherington and Ranson (Anonymous 1983) demonstrated that lesions of the ventromedial hypothalamus (VMH) lead to obesity in rats, while in the 1950s, experiments introducing lesions into the lateral hypothalamus

C. Schulz, K. Paulus, and H. Lehnert (✉)
Department of Internal Medicine I, Luebeck University, Ratzeburger Allee 160, 23538 Luebeck, Germany
e-mail: hendrik.lehnert@uk-h.de

W. Meyerhof et al. (eds.), *Sensory and Metabolic Control of Energy Balance*, Results and Problems in Cell Differentiation 52,
DOI 10.1007/978-3-642-14426-4_16, 

(LHA) elicited a reduction in food intake and loss of body weight (Anand and Brobeck 1951).

The existence of one or more peripheral signals reflecting the body's energy stores and feeding this information to the hypothalamus was proposed (Brobeck 1946; Mayer 1955) and it was Kennedy who suggested that an adipose tissue–derived factor might be involved (Kennedy 1953). This hypothesis was supported by Hervey's parabiosis experiments with VMH-lesioned (and thus obese) rats paired with intact rats. In these pairs, the intact rat lost weight, suggesting that the lesioned animal overproduces a circulating factor that reduces food intake and body weight in the intact rat (Hervey 1959). However, the underlying agent remained unknown.

Other parabiosis experiments involving obese mice of the db and the ob strain were performed by Coleman. In his study, ob and db mice were paired with each other and with wild-type mice. The results, namely weight loss in wild-type and ob mice when paired with a db animal, led to the conclusion that the product of the ob gene is a circulating factor that suppresses food intake and reduces body weight and that the db gene encodes for the receptor of this factor (Coleman 1973).

However, it was only in 1994, when the ob gene was identified by positional cloning (Zhang et al. 1994) as a ca. 4.5 kb RNA encoding for a 167 amino acid polypeptide with a signal sequence identifying it as a secreted peptide. This peptide was named leptin (from the Greek root leptos, meaning "thin") and it was shown to serve as an afferent signal in a negative feedback loop that maintains stability of the adipose tissue mass (Zhang et al. 1994; Halaas et al. 1995; Pelleymounter et al. 1995; Campfield et al. 1995). In fact, leptin concentrations are positively correlated with the amount of body fat (Brennan and Mantzoros 2006, 2007; Blüher and Mantzoros 2007; Chan et al. 2006; Kelesidis and Mantzoros 2006; Chan and Mantzoros 2005; Lee et al. 2006). Soon after the cloning of the leptin gene, the leptin receptor (obR) with its several splice variants was discovered (Tartaglia et al. 1995; Chen et al. 1996; Lee et al. 1996).

To access the brain, leptin is transported across the blood–brain barrier (BBB) via a saturable mechanism involving the short splice variant of the leptin receptor (obRa) and through a second, not yet identified, transport mechanism (Banks et al. 1996). Other potential points of entry are the circumventricular organs (CVOs), e.g., the arcuate nucleus (ARC), the subfornical organ, and area postrema. The CVOs are unique in that they are extensively vascularized and possess highly fenestrated capillaries. This neurovascular interface controls the penetration of molecules into the brain (Begley 1994; Strand 1999) and has been shown to be permeable to leptin (Ahima et al. 2000).

Within the central nervous system (CNS), leptin binds to the long form of the leptin receptor, obRb, thereby reducing food intake and increasing energy expenditure, together resulting in a reduction of body weight. Within the brain, the highest levels of the obRb are present in several hypothalamic nuclei, including the ARC, the dorsomedial (DMH), the VMH, the LHA, and the ventral premammillary (PMV) nuclei (Elmquist et al. 1998, 1999; Baskin et al. 1999; Leshan et al. 2006). Furthermore, the obRb is found in the ventral tegmental area (VTA), in the brainstem (including the nucleus of the solitary tract (NTS) and the dorsal motor nucleus of the vagus) and in the periaqueductal grey matter, among others.

To our current knowledge, the ARC is a major target for leptin. In this nucleus, the obRb is present on two different populations of neurons, one of which is synthesizing neuropeptide Y (NPY) and agouti-related peptide (Agrp) and is located medially (Elmquist et al. 1999; Schwartz et al. 2000) and a second population of neurons expressing proopiomelanocortin (POMC) and cocaine and amphetamine-regulated transcript (CART), which is found laterally. The orexigenic NPY/Agrp neurons are inhibited by leptin, while the anorexigenic POMC/CART neurons are activated (Ahima et al. 2000; Jobst et al. 2004; Myers et al. 2008). POMC is processed to produce α-melanocyte-stimulating hormone (α-MSH) and adrenocorticotropic hormone (ACTH), both of which act via melanocortin 3 and 4 (MC3, MC4) receptors to induce satiety (Schulz et al. 2009; Huszar et al. 1997; Butler and Cone 2002; Marsh et al. 1999; Butler et al. 2000; Chen et al. 2000). Agrp is an antagonist of α-MSH/ACTH signaling at the MC4 (Schwartz 2006; Ollmann et al. 1997).

Upon activation of the obRb by leptin, the receptor-associated Janus kinase 2 (JAK2) becomes activated by auto or cross phosphorylation and phosphorylates tyrosine residues 985 and 1138 in the cytoplasmic domain of the receptor, followed by phosphorylation and activation of the signal transducer and activator of transcription (STAT3). Activated STAT3 dimerizes, translocates to the nucleus, and activates its target genes with inducing their expression, including the suppressor of cytokine signaling 3 (SOCS3). SOCS3 takes part in a feedback loop that inhibits leptin signaling by binding to phosphorylated tyrosines of JAK2. Other adaptor proteins are recruited to activate phosphatidylinositol-3 phosphate kinase (PI3K) and extracellular signal-regulated kinase 1/2 (ERK1/2). Dephosphorylation of the JAK2 leads to internalization of the obRb. ERK1/2 can downregulate a protein tyrosine phosphatase, thereby indirectly increasing STAT3 activity. PI3K enhances the activity of phosphodiesterase 3B (PDE3B), which increases AMP formation from cAMP and thereby induces adenosine monophosphate kinase (AMPK). AMPK activity can also be enhanced by JAK2- and STAT3-dependent signaling directly. AMPK in turn can induce and orchestrate peroxisome proliferator–activated receptor $\gamma$ coactivator and peroxisome proliferator–activated receptor signaling (Poeggeler et al. 2009).

As pointed out above, leptin concentrations are positively correlated with the amount of body fat (Brennan and Mantzoros 2006, 2007; Blüher and Mantzoros 2007; Chan et al. 2006; Kelesidis and Mantzoros 2006; Chan and Mantzoros 2005; Lee et al. 2006), but although obese experimental animals and humans are hyperleptinemic compared to lean subjects, they are less sensitive to leptin. This is commonly referred to as leptin resistance (Brennan and Mantzoros 2006; Chan and Mantzoros 2005). It is thought that leptin resistance results from alterations in obR signaling in hypothalamic neurons (particularly the ARC) and/or transport across the BBB (Myers et al. 2008; Münzberg 2008). The latter can be circumvented by intranasal application of leptin. We have shown that leptin reaches the hypothalamus in supraphysiological amounts via this route (Fliedner et al. 2006) and that it acts to reduce food intake and body weight gain in rats (Schulz et al. 2004).

The discovery of leptin initiated an increased exploration into the endocrine potential of adipose tissue and the quest to elucidate the crosstalk between adipose tissue and other organs. Consequently, numerous factors synthesized and released by white adipose tissue (WAT) have been identified and the term "adipokines" was coined. They are generally defined as biologically active substances produced in adipose tissue that act in an autocrine, paracrine, or endocrine fashion. Comparable to leptin, some adipokines were identified to be involved in body weight homeostasis.

Adiponectin is the most abundant secreted protein produced by the WAT (Kadowaki and Yamauchi 2005; Yildiz et al. 2004) and circulates in plasma in picogram to nanogram per milliliter concentrations in the form of homotrimers, low-molecular weight hexamers and high-molecular weight (HMW) complexes (Kadowaki et al. 2008). So far, it has not been clearly established which forms of adiponectin are biologically active, but based on clinical observations, current consensus is that the HMW form is most relevant (Dridi and Taouis 2009). In contrast to leptin, adiponectin is decreased in obesity, is inversely related to glucose and insulin, and increases during fasting (Kadowaki and Yamauchi 2005). It is suggestive that adiponectin contributes to the crosstalk between adipose tissue and the brain.

Adiponectin deficiency induces insulin resistance and hyperlipidemia and increases the susceptibility toward vascular injury and atherosclerosis (Kadowaki et al. 2008). The administration of insulin-sensitizing thiazolidinediones increases HMW adiponectin in both humans and rodents (Bodles et al. 2006; Tsuchida et al. 2005).

There are two forms of adiponectin receptors, AdipoR1 and AdipoR2, that are expressed in many tissues, including the CNS (Kadowaki et al. 2008). Both receptors are widely distributed throughout the brain, notably in the hypothalamus and the brainstem (Wilkinson et al. 2007; Kubota et al. 2007; Fry et al. 2006; Kos et al. 2007). Immunohistochemical analysis revealed colocalization of the AdipoR1 with obRb in the ARC of mice (Kubota et al. 2007), and it was shown that AdipoR1 and AdipoR2 are present in both POMC/CART and NPY/Agrp neurons (Guillod-Maximin et al. 2009).

Trimeric and low-molecular weight adiponectin have been detected in the cerebrospinal fluid (CSF) of humans and rodents; both increase following i.v. application (Kubota et al. 2007; Kusminski et al. 2007; Qi et al. 2004). The lack of HMW (>500 kDa) in CSF implicates that only smaller forms can pass through the BBB (Kubota et al. 2007; Kusminski et al. 2007). However, there are conflicting data about the transport of adiponectin from the periphery into the CNS and whether this transport actually takes place remains a matter of debate (Qi et al. 2004; Pan and Kastin 2007; Pan et al. 2006; Spranger et al. 2006). Assuming that this transport exists, it is likely that adiponectin enters the brain via receptor-mediated transcytosis (Kos et al. 2007). Another potential way is via the CVOs (Fry et al. 2006; Ahima et al. 2006). However, the CVOs have a total surface area of only 0.02 $cm^2/g$ tissue by contrast to the surface area of the BBB of 100–150 $cm^2/g$

(Begley 1994; Strand 1999), rendering it unlikely that amounts of adiponectin, which become detectable in CSF, can enter the brain via this route. Adiponectin in the CSF might also be derived directly from the CNS: although the WAT is by far the largest source of adiponectin, it is also produced within the CNS, as was shown in rodents and chicken (Wilkinson et al. 2007; Maddineni et al. 2005).

A number of experiments indicate that adiponectin exerts central nervous actions, particularly with respect to energy homeostasis, but there are conflicting data and further studies are needed to clarify adiponectin's CNS actions. One study has shown that adiponectin i.c.v. increases energy expenditure, but has no effect on food intake (Qi et al. 2004). In other publications, adiponectin reduced food intake (Coope et al. 2008; Shklyaev et al. 2003) via AdipoR1 (Coope et al. 2008); the latter was accompanied by an activation of IRS1/2, ERK, Akt, FOXO1, JAK2, and STAT3, indicating common mechanisms or an interaction with insulin and leptin signaling (Coope et al. 2008). Peripheral administration of adiponectin reduces body weight by enhancing fatty acid oxidation, but has no apparent effects on food intake (Yamauchi et al. 2001; Tomas et al. 2002; Berg et al. 2001). However, Kubota et al. reported that peripheral application of adiponectin rather exerted an orexigenic effect and decreased energy expenditure (Kubota et al. 2007).

NEFA/nucleobindin 2–encoded satiety- and fat-influencing protein (nesfatin) is an adipokine that has recently come into focus for its anorexigenic effects. It is processed from the precursor peptide NEFA/nucleobindin 2 (NUCB2) that is expressed in the CNS, in WAT as well as in other peripheral organs, e.g., gastric mucosa and pancreatic β-cells (Shimizu et al. 2009a; Ramanjaneya 2010; Stengel et al. 2009a; Gonzalez et al. 2009), and it was found in 3T3-L1 adipocytes (Adachi et al. 2007). By Western blot, it was detected in rat serum and CSF (Shimizu et al. 2009a). NUCB2/nesfatin was originally described as a secreted protein of unknown function (Miura et al. 1992; Barnikol-Watanabe et al. 1994), but in 2005 and 2006, its anorexigenic properties were reported (Oh-I et al. 2005, 2006). In the CNS, NUCB2/nesfatin mRNA is present, e.g., in the hypothalamus (ARC, paraventricular nucleus (PVN), supraoptic nucleus (SON), LHA) and brainstem (e.g., NTS, dorsal motor nucleus of the vagus) (Oh-I et al. 2006; Brailoiu et al. 2007).

In the PVN and the SON, starvation reduces NUCB2/nesfatin mRNA expression, which is restored by refeeding (Oh-I et al. 2006; Kohno et al. 2008); other hypothalamic nuclei are unaffected. I.c.v. NUCB2/nesfatin dose dependently reduces food intake in rats (Oh-I et al. 2006). NUCB2/Nesfatin possesses potential cleaving sites for protein convertases PC1/2 and PC3 and is colocalized with these enzymes in the cytoplasm (Oh-I et al. 2006; Steiner et al. 1992), suggesting that further processing of the NUCB2/nesfatin might take place physiologically. This is also supported by the detection of nesfatin-1, the amino terminal fragment of NUCB2/nesfatin, in the CSF (Oh-I et al. 2006).

Consequently, fragments of NUCB2/nesfatin were examined in animal experiments and it turned out that nesfatin-1 exerts an anorexigenic effect, which is dose dependent, while other fragments tested were ineffective in terms of food intake (Oh-I et al. 2006). Furthermore, immunoblockade of endogenous nesfatin-1 by i.c.v. antibody application stimulates food intake (Oh-I et al. 2006).

While nesfatin-1 is produced locally within the CNS, it has also been shown by two groups independently that it can cross the BBB via nonsaturable transmembrane diffusion, which is consistent with its low lipophilicity (Pan et al. 2007; Price et al. 2007). This raises the possibility that peripheral nesfatin-1 (either endogenous or exogenous) may access the CNS. In fact, i.p. or s.c. application of nesfatin-1 inhibited food intake both acutely and chronically (Shimizu et al. 2009a, b; Oh-I et al. 2005). Most interestingly, like leptin (see above), intranasal application of nesfatin-1 reduces food intake in male Wistar rats (Shimizu et al. 2009a).

Within the hypothalamus, NUCB2/nesfatin interacts with the melanocortin system (Oh-I et al. 2006). It has been shown that central nervous application of α-MSH stimulates NUCB2/nesfatin expression in the PVN. Furthermore, nesfatin-1-induced anorexia can be abolished by blocking of the MC3 and MC4 receptors (Oh-I et al. 2006). However, it is debatable that the anorexigenic activity of nesfatin-1 is based on a direct agonistic action at the MCR, since nesfatin-1 does not elicit an increase of intracellular cAMP, which is characteristic for G-protein–coupled receptors such as the MCRs, in an in vitro system (Oh-I et al. 2006). Activation of MC3 and MC4, however, can also lead to an increase in intracellular $Ca^{2+}$ (Mountjoy et al. 2001) and Brailoiu et al. reported that nesfatin-1 interacts with a G-protein-coupled receptor to increase $Ca^{2+}$ in cultured hypothalamic neurons; whether this receptor is one of the MCRs, however, remains to be determined (Brailoiu et al. 2007).

Nesfatin-1 affects the excitability of a large proportion of different subpopulations of neurons located in the PVN (Price et al. 2008) and both in this nucleus and in the SON, NUCB2/nesfatin is colocalized with vasopressin and oxytocin (Kohno et al. 2008). In the PVN, NUCB2/nesfatin is found in the same neurons as thyrotropin-releasing hormone (TRH), corticotropin-releasing factor (CRF), somatostatin, neurotensin, and growth-releasing hormone (Douglas et al. 2007). The physiological importance of NUCB2/nesfatin interactions with CRF is supported by a recent publication on the relevance of the $CRF_2$ receptor for nesfatin-1's actions in the forebrain (Stengel et al. 2009b).

NUCB2/nesfatin immunoreactivity is also found in POMC/CART neurons of the ARC (Oh-I et al. 2006; Brailoiu et al. 2007) and in NPY/Agrp neurons (Inhoff et al. 2009). The latter neurons are hyperpolarized by nesfatin-1 (Price et al. 2008). Furthermore, NUCB2/nesfatin is coexpressed with the anorexigenic neuropeptide melanin-concentrating hormone (MCH) in tuberal hypothalamic neurons of the rat (Fort et al. 2008).

In 2005, by employing the method of differential display of expressed genes in paired samples of subcutaneous and visceral fat from two female volunteers, an adipokine primarily synthesized by visceral fat was discovered and named "visfatin" (Fukuhara et al. 2005). Subsequently, it turned out that this molecule had previously been described as "pre-B-cell colony-enhancing factor" (PBEF) in liver, skeletal muscle, and bone marrow as a growth factor for B-lymphocyte precursors (Samal et al. 1994; Jia et al. 2004). It is also known as nicotinamide phosphoribosyl transferase (Nampt), since it is acting as an enzyme catalyzing the condensation of nicotinamide with phosphoribosyl pyrophosphate, representing the first step in the salvage pathway allowing recycling of nicotinamide to NAD (Rongvaux et al. 2002, 2008; Revollo et al. 2004).

Visfatin is present in both human and murine plasma and is strongly correlated with fat mass (Fukuhara et al. 2005; Filippatos et al. 2008; Berndt et al. 2005), although the primary amino acid sequence of visfatin does not include a signal sequence, as would be expected in secreted proteins. Other researchers have detected visfatin rather in the cell nucleus and cytoplasm (Anonymous 1983; Rongvaux et al. 2002; Kitani et al. 2003). Recently, these apparent contradictions were reconciled by demonstrating that visfatin is secreted by adipocytes via a nonclassical secretory pathway, which is highly regulated (Revollo et al. 2007; Tanaka et al. 2007).

In normal mice, i.v. infusion of visfatin leads to an acute decline in plasma glucose, which is independent of insulin secretion (Fukuhara et al. 2005); mouse models of type II diabetes also exhibit a decrease in plasma glucose (Fukuhara et al. 2005). Chronic supplementation with visfatin (by adenoviral vector carrying the visfatin gene) results in a slight decrease of plasma glucose and insulin (Fukuhara et al. 2005). Visfatin, like insulin, stimulates glucose uptake by cultured adipocytes and muscle cells and suppresses glucose release by cultured hepatocytes by binding a specific subtype of the insulin receptor and eliciting downstream signaling events, but does not compete with insulin (Fukuhara et al. 2005). However, these findings recently have come into doubt and still little is known about the function of this adipokine in energy homeostasis (Revollo et al. 2007; Fukuhara et al. 2007). Interestingly, in obese OLETF rats, visfatin mRNA in visceral fat is increased upon treatment with the insulin-sensitizing thiazlidinedione rosiglitazone (Choi et al. 2005).

Although visfatin is known for some time now, only recently a potential role in the CNS in the context of body weight homeostasis was suggested. We were able to show that visfatin is present in human CSF and that a negative relationship between CSF visfatin levels and adiposity exists, while plasma visfatin and body fat mass, body weight, and waist and hip circumference are positively correlated (Hallschmid et al. 2009). This suggests that visfatin's transport into the brain/CSF is impaired in obesity. Though it is not yet experimentally tested, the existing relationship of visfatin between CSF and plasma suggests that CSF visfatin is derived from the periphery; however, further studies are needed to support this finding. Since at least in dogs, visfatin is also produced within the CNS (McGlothlin et al. 2005), it is also possible that CSF visfatin is of central nervous origin and that CNS secretion is controlled by a peripheral agent, e.g., another adipokine, feeding back the status of adipose tissue. At present, the central nervous effects of visfatin are not explored in detail. To our knowledge, there is only one study in chicken, in which visfatin exerted an orexigenic effect after i.c.v. administration and affected c-Fos immunoreactivity differentially in distinct hypothalamic nuclei (Cline et al. 2008).

With this article, we could only address a few of the steadily increasing number of known adipokines. As an ongoing stream, new substances secreted by WAT are identified and some of them, in the process of research, turn out to be involved in the crosstalk between adipose tissue and the CNS. The exploration of the endocrine potential of the adipose tissue remains an ongoing quest for researchers and exciting discoveries will continue to be made.

## References


Adachi J, Kumar C, Zhang Y, Mann M (2007) In-depth analysis of the adipocyte proteome by mass spectrometry and bioinformatics. Mol Cell Proteomics 6:1257–1273

Ahima RS, Saper CB, Flier JS, Elmquist JK (2000) Leptin regulation of neuroendocrine systems. Front Neuroendocrinol 21:263–307

Ahima RS, Qi Y, Singhal NS, Jackson MB, Scherer PE (2006) Brain adipocytokine action and metabolic regulation. Diabetes 55(Suppl 2):S145–S154

Anand BK, Brobeck JR (1951) Hypothalamic control of food intake in rats and cats. Yale J Biol Med 24:123–140

Anonymous (1983) Nutrition classics. The anatomical record, volume 78, 1940: hypothalamic lesions and adiposity in the rat 1983. Nutr Rev 41:124–127

Banks WA, Kastin AJ, Huang W, Jaspan JB, Maness LM (1996) Leptin enters the brain by a saturable system independent of insulin. Peptides 17:305–311

Barnikol-Watanabe S, Gross NA, Gotz H, Henkel T, Karabinos A, Kratzin H, Barnikol HU, Hilschmann N (1994) Human protein NEFA, a novel DNA binding/EF-hand/leucine zipper protein. Molecular cloning and sequence analysis of the cDNA, isolation and characterization of the protein. Biol Chem Hoppe Seyler 375:497–512

Baskin DG, Schwartz MW, Seeley RJ, Woods SC, D P Jr, Breininger JF, Jonak Z, Schaefer J, Krouse M, Burghardt C, Campfield LA, Burn P, Kochan JP (1999) Leptin receptor long-form splice-variant protein expression in neuron cell bodies of the brain and co-localization with neuropeptide Y mRNA in the arcuate nucleus. J Histochem Cytochem 47:353–362

Begley DJ (1994) Peptides and the blood–brain barrier: the status of our understanding. Ann N Y Acad Sci 739:89–100

Berg AH, Combs TP, Du X, Brownlee M, Scherer PE (2001) The adipocyte-secreted protein Acrp30 enhances hepatic insulin action. Nat Med 7:947–953

Berndt J, Kloting N, Kralisch S, Kovacs P, Fasshauer M, Schon MR, Stumvoll M, Blüher M (2005) Plasma visfatin concentrations and fat depot-specific mRNA expression in humans. Diabetes 54:2911–2916

Blüher S, Mantzoros CS (2007) Leptin in reproduction. Curr Opin Endocrinol Diabetes Obes 14:458–464

Bodles AM, Banga A, Rasouli N, Ono F, Kern PA, Owens RJ (2006) Pioglitazone increases secretion of high-molecular-weight adiponectin from adipocytes. Am J Physiol Endocrinol Metab 291:E1100–E1105

Brailoiu GC, Dun SL, Brailoiu E, Inan S, Yang J, Chang JK, Dun NJ (2007) Nesfatin-1: distribution and interaction with a G protein-coupled receptor in the rat brain. Endocrinology 148:5088–5094

Brennan AM, Mantzoros CS (2006) Drug insight: the role of leptin in human physiology and pathophysiology – emerging clinical applications. Nat Clin Pract Endocrinol Metab 2:318–327

Brennan AM, Mantzoros CS (2007) Leptin and adiponectin: their role in diabetes. Curr Diab Rep 7:1–2

Brobeck JR (1946) Mechanism of the development of obesity in animals with hypothalamic lesions. Physiol Rev 25:541–559

Butler AA, Cone RD (2002) The melanocortin receptors: lessons from knockout models. Neuropeptides 36:77–84

Butler AA, Kesterson RA, Khong K, Cullen MJ, Pelleymounter MA, Dekoning J, Baetscher M, Cone RD (2000) A unique metabolic syndrome causes obesity in the melanocortin-3 receptor-deficient mouse. Endocrinology 141:3518–3521

Campfield LA, Smith FJ, Guisez Y, Devos R, Burn P (1995) Recombinant mouse OB protein: evidence for a peripheral signal linking adiposity and central neural networks. Science 269:546–549

Chan JL, Mantzoros CS (2005) Role of leptin in energy-deprivation states: normal human physiology and clinical implications for hypothalamic amenorrhoea and anorexia nervosa. Lancet 366:74–85

Chan JL, Matarese G, Shetty GK, Raciti P, Kelesidis I, Aufiero D, De Rosa V, Perna F, Fontana S, Mantzoros CS (2006) Differential regulation of metabolic, neuroendocrine, and immune function by leptin in humans. Proc Natl Acad Sci USA 103:8481–8486

Chen H, Charlat O, Tartaglia LA, Woolf EA, Weng X, Ellis SJ, Lakey ND, Culpepper J, Moore KJ, Breitbart RE, Duyk GM, Tepper RI, Morgenstern JP (1996) Evidence that the diabetes gene encodes the leptin receptor: identification of a mutation in the leptin receptor gene in db/db mice. Cell 84:491–495

Chen AS, Marsh DJ, Trumbauer ME, Frazier EG, Guan XM, Yu H, Rosenblum CI, Vongs A, Feng Y, Cao L, Metzger JM, Strack AM, Camacho RE, Mellin TN, Nunes CN, Min W, Fisher J, Gopal-Truter S, MacIntyre DE, Chen HY, Van der Ploeg LH (2000) Inactivation of the mouse melanocortin-3 receptor results in increased fat mass and reduced lean body mass. Nat Genet 26:97–102

Choi KC, Ryu OH, Lee KW, Kim HY, Seo JA, Kim SG, Kim NH, Choi DS, Baik SH, Choi KM (2005) Effect of PPAR-alpha and -gamma agonist on the expression of visfatin, adiponectin, and TNF-alpha in visceral fat of OLETF rats. Biochem Biophys Res Commun 336:747–753

Cline MA, Nandar W, Prall BC, Bowden CN, Denbow DM (2008) Central visfatin causes orexigenic effects in chicks. Behav Brain Res 186:293–297

Coleman DL (1973) Effects of parabiosis of obese with diabetes and normal mice. Diabetologia 9:294–298

Coope A, Milanski M, Araujo EP, Tambascia M, Saad MJ, Geloneze B, Velloso LA (2008) AdipoR1 mediates the anorexigenic and insulin/leptin-like actions of adiponectin in the hypothalamus. FEBS Lett 582:1471–1476

Douglas AJ, Johnstone LE, Leng G (2007) Neuroendocrine mechanisms of change in food intake during pregnancy: a potential role for brain oxytocin. Physiol Behav 91:352–365

Dridi S, Taouis M (2009) Adiponectin and energy homeostasis: consensus and controversy. J Nutr Biochem 20:831–839

Elmquist JK, Bjorbaek C, Ahima RS, Flier JS, Saper CB (1998) Distributions of leptin receptor mRNA isoforms in the rat brain. J Comp Neurol 395:535–547

Elmquist JK, Elias CF, Saper CB (1999) From lesions to leptin: hypothalamic control of food intake and body weight. Neuron 22:221–232

Filippatos TD, Derdemezis CS, Gazi IF, Lagos K, Kiortsis DN, Tselepis AD, Elisaf MS (2008) Increased plasma visfatin levels in subjects with the metabolic syndrome. Eur J Clin Invest 38:71–72

Fliedner S, Schulz C, Lehnert H (2006) Brain uptake of intranasally applied radioiodinated leptin in wistar rats. Endocrinology 147:2088–2094

Fort P, Salvert D, Hanriot L, Jego S, Shimizu H, Hashimoto K, Mori M, Luppi PH (2008) The satiety molecule nesfatin-1 is co-expressed with melanin concentrating hormone in tuberal hypothalamic neurons of the rat. Neuroscience 155:174–181

Fry M, Smith PM, Hoyda TD, Duncan M, Ahima RS, Sharkey KA, Ferguson AV (2006) Area postrema neurons are modulated by the adipocyte hormone adiponectin. J Neurosci 20;26:9695–9702

Fukuhara A, Matsuda M, Nishizawa M, Segawa K, Tanaka M, Kishimoto K, Matsuki Y, Murakami M, Ichisaka T, Murakami H, Watanabe E, Takagi T, Akiyoshi M, Ohtsubo T, Kihara S, Yamashita S, Makishima M, Funahashi T, Yamanaka S, Hiramatsu R, Matsuzawa Y, Shimomura I (2005) Visfatin: a protein secreted by visceral fat that mimics the effects of insulin. Science 307:426–430

Fukuhara A, Matsuda M, Nishizawa M, Segawa K, Tanaka M, Kishimoto K, Matsuki Y, Murakami M, Ichisaka T, Murakami H, Watanabe E, Takagi T, Akiyoshi M, Ohtsubo T, Kihara S, Yamashita S, Makishima M, Funahashi T, Yamanaka S, Hiramatsu R, Matsuzawa Y, Shimomura I (2007) Retraction. Science 318:565

Gonzalez R, Tiwari A, Unniappan S (2009) Pancreatic beta cells colocalize insulin and pronesfatin immunoreactivity in rodents. Biochem Biophys Res Commun 381:643–648

Guillod-Maximin E, Roy AF, Vacher CM, Aubourg A, Bailleux V, Lorsignol A, Penicaud L, Parquet M, Taouis M (2009) Adiponectin receptors are expressed in hypothalamus and colocalized with proopiomelanocortin and neuropeptide Y in rodent arcuate neurons. J Endocrinol 200:93–105

Halaas JL, Gajiwala KS, Maffei M, Cohen SL, Chait BT, Rabinowitz D, Lallone RL, Burley SK, Friedman JM (1995) Weight-reducing effects of the plasma protein encoded by the obese gene. Science 269:543–546

Hallschmid M, Randeva H, Tan BK, Kern W, Lehnert H (2009) Relationship between cerebrospinal fluid visfatin (PBEF/Nampt) levels and adiposity in humans. Diabetes 58:637–640

Hervey GR (1959) The effects of lesions in the hypothalamus in parabiotic rats. J Physiol 145:336–352

Huszar D, Lynch CA, Fairchild-Huntress V, Dunmore JH, Fang Q, Berkemeier LR, Gu W, Kesterson RA, Boston BA, Cone RD, Smith FJ, Campfield LA, Burn P, Lee F (1997) Targeted disruption of the melanocortin-4 receptor results in obesity in mice. Cell 88:131–141

Inhoff T, Stengel A, Peter L, Goebel M, Tache Y, Bannert N, Wiedenmann B, Klapp BF, Monnikes H, Kobelt P (2009) Novel insight in distribution of nesfatin-1 and phospho-mTOR in the arcuate nucleus of the hypothalamus of rats. Peptides 31:257–262

Jia SH, Li Y, Parodo J, Kapus A, Fan L, Rotstein OD, Marshall JC (2004) Pre-B cell colony-enhancing factor inhibits neutrophil apoptosis in experimental inflammation and clinical sepsis. J Clin Invest 113:1318–1327

Jobst EE, Enriori PJ, Cowley MA (2004) The electrophysiology of feeding circuits. Trends Endocrinol Metab 15:488–499

Kadowaki T, Yamauchi T (2005) Adiponectin and adiponectin receptors. Endocr Rev 26:439–451

Kadowaki T, Yamauchi T, Kubota N (2008) The physiological and pathophysiological role of adiponectin and adiponectin receptors in the peripheral tissues and CNS. FEBS Lett 582:74–80

Kelesidis T, Mantzoros CS (2006) The emerging role of leptin in humans. Pediatr Endocrinol Rev 3:239–248

Kennedy GC (1953) The role of depot fat in the hypothalamic control of food intake in the rat. Proc R Soc Lond B Biol Sci 140:578–596

Kitani T, Okuno S, Fujisawa H (2003) Growth phase-dependent changes in the subcellular localization of pre-B-cell colony-enhancing factor. FEBS Lett 544:74–78

Kohno D, Nakata M, Maejima Y, Shimizu H, Sedbazar U, Yoshida N, Dezaki K, Onaka T, Mori M, Yada T (2008) Nesfatin-1 neurons in paraventricular and supraoptic nuclei of the rat hypothalamus coexpress oxytocin and vasopressin and are activated by refeeding. Endocrinology 149:1295–1301

Kos K, Harte AL, da Silva NF, Tonchev A, Chaldakov G, James S, Snead DR, Hoggart B, O'Hare JP, McTernan PG, Kumar S (2007) Adiponectin and resistin in human cerebrospinal fluid and expression of adiponectin receptors in the human hypothalamus. J Clin Endocrinol Metab 92:1129–1136

Kubota N, Yano W, Kubota T, Yamauchi T, Itoh S, Kumagai H, Kozono H, Takamoto I, Okamoto S, Shiuchi T, Suzuki R, Satoh H, Tsuchida A, Moroi M, Sugi K, Noda T, Ebinuma H, Ueta Y, Kondo T, Araki E, Ezaki O, Nagai R, Tobe K, Terauchi Y, Ueki K, Minokoshi Y, Kadowaki T (2007) Adiponectin stimulates AMP-activated protein kinase in the hypothalamus and increases food intake. Cell Metab 6:55–68

Kusminski CM, McTernan PG, Schraw T, Kos K, O'Hare JP, Ahima R, Kumar S, Scherer PE (2007) Adiponectin complexes in human cerebrospinal fluid: distinct complex distribution from serum. Diabetologia 50:634–642

Lee GH, Proenca R, Montez JM, Carroll KM, Darvishzadeh JG, Lee JI, Friedman JM (1996) Abnormal splicing of the leptin receptor in diabetic mice. Nature 379:632–635

Lee JH, Chan JL, Sourlas E, Raptopoulos V, Mantzoros CS (2006) Recombinant methionyl human leptin therapy in replacement doses improves insulin resistance and metabolic profile in patients with lipoatrophy and metabolic syndrome induced by the highly active antiretroviral therapy. J Clin Endocrinol Metab 91:2605–2611

Leshan RL, Bjornholm M, Münzberg H, Myers MG Jr (2006) Leptin receptor signaling and action in the central nervous system. Obesity (Silver Spring) 14(Suppl 5):208S–212S

Maddineni S, Metzger S, Ocon O, Hendricks G III, Ramachandran R (2005) Adiponectin gene is expressed in multiple tissues in the chicken: food deprivation influences adiponectin messenger ribonucleic acid expression. Endocrinology 146:4250–4256

Marsh DJ, Hollopeter G, Huszar D, Laufer R, Yagaloff KA, Fisher SL, Burn P, Palmiter RD (1999) Response of melanocortin-4 receptor-deficient mice to anorectic and orexigenic peptides. Nat Genet 21:119–122

Mayer J (1955) Regulation of energy intake and the body weight: the glucostatic theory and the lipostatic hypothesis. Ann N Y Acad Sci 63:15–43

McGlothlin JR, Gao L, Lavoie T, Simon BA, Easley RB, Ma SF, Rumala BB, Garcia JG, Ye SQ (2005) Molecular cloning and characterization of canine pre-B-cell colony-enhancing factor. Biochem Genet 43:127–141

Miura K, Titani K, Kurosawa Y, Kanai Y (1992) Molecular cloning of nucleobindin, a novel DNA-binding protein that contains both a signal peptide and a leucine zipper structure. Biochem Biophys Res Commun 187:375–380

Mountjoy KG, Kong PL, Taylor JA, Willard DH, Wilkison WO (2001) Melanocortin receptor-mediated mobilization of intracellular free calcium in HEK293 cells. Physiol Genomics 5:11–19

Münzberg H (2008) Differential leptin access into the brain – a hierarchical organization of hypothalamic leptin target sites? Physiol Behav 94:664–669

Myers MG, Cowley MA, Münzberg H (2008) Mechanisms of leptin action and leptin resistance. Annu Rev Physiol 70:537–556

Oh-I S, Shimizu H, Sato T, Uehara Y, Okada S, Mori M (2005) Molecular mechanisms associated with leptin resistance: n-3 polyunsaturated fatty acids induce alterations in the tight junction of the brain. Cell Metab 1:331–341

Oh-I S, Shimizu H, Satoh T, Okada S, Adachi S, Inoue K, Eguchi H, Yamamoto M, Imaki T, Hashimoto K, Tsuchiya T, Monden T, Horiguchi K, Yamada M, Mori M (2006) Identification of nesfatin-1 as a satiety molecule in the hypothalamus. Nature 443:709–712

Ollmann MM, Wilson BD, Yang YK, Kerns JA, Chen Y, Gantz I, Barsh GS (1997) Antagonism of central melanocortin receptors in vitro and in vivo by agouti-related protein. Science 278:135–138

Pan W, Kastin AJ (2007) Adipokines and the blood–brain barrier. Peptides 28:1317–1330

Pan W, Tu H, Kastin AJ (2006) Differential BBB interactions of three ingestive peptides: obestatin, ghrelin, and adiponectin. Peptides 27:911–916

Pan W, Hsuchou H, Kastin AJ (2007) Nesfatin-1 crosses the blood–brain barrier without saturation. Peptides 28:2223–2228

Pelleymounter MA, Cullen MJ, Baker MB, Hecht R, Winters D, Boone T, Collins F (1995) Effects of the obese gene product on body weight regulation in ob/ob mice [see comments]. Science 269:540–543

Poeggeler B, Schulz C, Pappolla MA, Bodo E, Tiede S, Lehnert H, Paus R (2009) Leptin and the skin: a new frontier. Exp Dermatol (in press)

Price TO, Samson WK, Niehoff ML, Banks WA (2007) Permeability of the blood–brain barrier to a novel satiety molecule nesfatin-1. Peptides 28:2372–2381

Price CJ, Hoyda TD, Samson WK, Ferguson AV (2008) Nesfatin-1 influences the excitability of paraventricular nucleus neurones. J Neuroendocrinol 20:245–250

Qi Y, Takahashi N, Hileman SM, Patel HR, Berg AH, Pajvani UB, Scherer PE, Ahima RS (2004) Adiponectin acts in the brain to decrease body weight. Nat Med 10:524–529

Ramanjaneya M, Chen J, Brown J, Patel S, Tan B, Randeva H (2010) Identification of nesfatin-1/NUCB2 as a novel depot-specific adipokine in human and murine adipose tissue: altered levels in obesity and food deprivation. pp 19, 131 (Abstract)
Revollo JR, Grimm AA, Imai S (2004) The NAD biosynthesis pathway mediated by nicotinamide phosphoribosyltransferase regulates Sir2 activity in mammalian cells. J Biol Chem 279:50754–50763
Revollo JR, Korner A, Mills KF, Satoh A, Wang T, Garten A, Dasgupta B, Sasaki Y, Wolberger C, Townsend RR, Milbrandt J, Kiess W, Imai S (2007) Nampt/PBEF/Visfatin regulates insulin secretion in beta cells as a systemic NAD biosynthetic enzyme. Cell Metab 6:363–375
Rongvaux A, Shea RJ, Mulks MH, Gigot D, Urbain J, Leo O, Andris F (2002) Pre-B-cell colony-enhancing factor, whose expression is up-regulated in activated lymphocytes, is a nicotinamide phosphoribosyltransferase, a cytosolic enzyme involved in NAD biosynthesis. Eur J Immunol 32:3225–3234
Rongvaux A, Galli M, Denanglaire S, Van Gool F, Dreze PL, Szpirer C, Bureau F, Andris F, Leo O (2008) Nicotinamide phosphoribosyl transferase/pre-B cell colony-enhancing factor/visfatin is required for lymphocyte development and cellular resistance to genotoxic stress. J Immunol 181:4685–4695
Samal B, Sun Y, Stearns G, Xie C, Suggs S, McNiece I (1994) Cloning and characterization of the cDNA encoding a novel human pre-B-cell colony-enhancing factor. Mol Cell Biol 14:1431–1437
Schulz C, Paulus K, Lehnert H (2004) Central nervous and metabolic effects of intranasally applied leptin. Endocrinology 145:2696–2701
Schulz C, Paulus K, Lobmann R, Dallman MF, Lehnert H (2009) Endogenous ACTH, not only {alpha}-melanocyte stimulating hormone, reduces food intake mediated by hypothalamic mechanisms. Am J Physiol Endocrinol Metab (in press)
Schwartz MW (2006) Central nervous system regulation of food intake. Obesity (Silver Spring) 14 (Suppl 1):1S–8S
Schwartz MW, Woods SC, D P Jr, Seeley RJ, Baskin DG (2000) Central nervous system control of food intake. Nature 404:661–671
Shimizu H, Oh I, Okada S, Mori M (2009a) Nesfatin-1: an overview and future clinical application. Endocr J 56:537–543
Shimizu H, Oh I, Hashimoto K, Nakata M, Yamamoto S, Yoshida N, Eguchi H, Kato I, Inoue K, Satoh T, Okada S, Yamada M, Yada T, Mori M (2009b) Peripheral administration of nesfatin-1 reduces food intake in mice: the leptin-independent mechanism. Endocrinology 150:662–671
Shklyaev S, Aslanidi G, Tennant M, Prima V, Kohlbrenner E, Kroutov V, Campbell-Thompson M, Crawford J, Shek EW, Scarpace PJ, Zolotukhin S (2003) Sustained peripheral expression of transgene adiponectin offsets the development of diet-induced obesity in rats. Proc Natl Acad Sci USA 100:14217–14222
Spranger J, Verma S, Gohring I, Bobbert T, Seifert J, Sindler AL, Pfeiffer A, Hileman SM, Tschop M, Banks WA (2006) Adiponectin does not cross the blood–brain barrier but modifies cytokine expression of brain endothelial cells. Diabetes 55:141–147
Steiner DF, Smeekens SP, Ohagi S, Chan SJ (1992) The new enzymology of precursor processing endoproteases. J Biol Chem 267:23435–23438
Stengel A, Goebel M, Yakubov I, Wang L, Witcher D, Coskun T, Tache Y, Sachs G, Lambrecht NW (2009a) Identification and characterization of nesfatin-1 immunoreactivity in endocrine cell types of the rat gastric oxyntic mucosa. Endocrinology 150:232–238
Stengel A, Goebel M, Wang L, Rivier J, Kobelt P, Monnikes H, Lambrecht NW, Tache Y (2009b) Central nesfatin-1 reduces dark-phase food intake and gastric emptying in rats: differential role of corticotropin-releasing factor2 receptor. Endocrinology 150:4911–4919
Strand FL (1999) Neuropeptides: regulators of physiological processes. MIT, Cambridge
Tanaka M, Nozaki M, Fukuhara A, Segawa K, Aoki N, Matsuda M, Komuro R, Shimomura I (2007) Visfatin is released from 3T3-L1 adipocytes via a non-classical pathway. Biochem Biophys Res Commun 359:194–201

Tartaglia LA, Dembski M, Weng X, Deng N, Culpepper J, Devos R, Richards GJ, Campfield LA, Clark FT, Deeds J (1995) Identification and expression cloning of a leptin receptor, OB-R. Cell 83:1263–1271

Tomas E, Tsao TS, Saha AK, Murrey HE, Zhang CC, Itani SI, Lodish HF, Ruderman NB (2002) Enhanced muscle fat oxidation and glucose transport by ACRP30 globular domain: acetyl-CoA carboxylase inhibition and AMP-activated protein kinase activation. Proc Natl Acad Sci USA 99:16309–16313

Tsuchida A, Yamauchi T, Takekawa S, Hada Y, Ito Y, Maki T, Kadowaki T (2005) Peroxisome proliferator-activated receptor (PPAR) alpha activation increases adiponectin receptors and reduces obesity-related inflammation in adipose tissue: comparison of activation of PPARalpha, PPARgamma, and their combination. Diabetes 54:3358–3370

Wilkinson M, Brown R, Imran SA, Ur E (2007) Adipokine gene expression in brain and pituitary gland. Neuroendocrinology 86:191–209

Yamauchi T, Kamon J, Waki H, Terauchi Y, Kubota N, Hara K, Mori Y, Ide T, Murakami K, Tsuboyama-Kasaoka N, Ezaki O, Akanuma Y, Gavrilova O, Vinson C, Reitman ML, Kagechika H, Shudo K, Yoda M, Nakano Y, Tobe K, Nagai R, Kimura S, Tomita M, Froguel P, Kadowaki T (2001) The fat-derived hormone adiponectin reverses insulin resistance associated with both lipoatrophy and obesity. Nat Med 7:941–946

Yildiz BO, Suchard MA, Wong ML, McCann SM, Licinio J (2004) Alterations in the dynamics of circulating ghrelin, adiponectin, and leptin in human obesity. Proc Natl Acad Sci USA 101:10434–10439

Zhang Y, Proenca R, Maffei M, Barone M, Leopold L, Friedman JM (1994) Positional cloning of the mouse obese gene and its human homologue. Nature 372:425–432

# Index

MIX
Papier aus verantwortungsvollen Quellen
Paper from responsible sources
FSC® C105338

If you have any concerns about our products,
you can contact us on
**ProductSafety@springernature.com**

In case Publisher is established outside the EU,
the EU authorized representative is:
**Springer Nature Customer Service Center GmbH**
**Europaplatz 3, 69115 Heidelberg, Germany**

Printed by Libri Plureos GmbH
in Hamburg, Germany